The Heavens Are Falling

The Scientific Prediction of Catastrophes in Our Time

The Heavens Are Falling

The Scientific Prediction of Catastrophes in Our Time

Walter J. Karplus

Plenum Press • New York and London

Library of Congress Cataloging-in-Publication Data

Karplus, Walter J.
    The heavens are falling : the scientific prediction of
  catastrophes in our time / Walter J. Karplus.
        p.   cm.
    Includes bibliographical references (p.   ) and index.
    ISBN 0-306-44130-6
    1. Disasters--Forecasting.   I. Title.
CB161.K36   1992
363.3'4'0112--dc20   91-44274
                                                            CIP

ISBN 0-306-44130-6

© 1992 American Interface Corporation
Plenum Press is a division of Plenum Publishing Corporation
233 Spring Street, New York, N.Y. 10013

Printed in the United States of America

To my wife, Takako,
and Maya and Tony

The Hen and the Heavens

ONCE upon a time a little red hen was picking up stones and worms and seeds in a barnyard when something fell on her head. "The heavens are falling down!" she shouted, and she began to run, still shouting, "The heavens are falling down!" All the hens that she met and all the roosters and turkeys and ducks laughed at her, smugly, the way you laugh at one who is terrified when you aren't. "What did you say?" they chortled. "The heavens are falling down!" cried the little red hen. Finally a very pompous rooster said to her, "Don't be silly, my dear, it was only a pea that fell on your head." And he laughed and laughed and everybody else except the little red hen laughed. Then suddenly with an awful roar great chunks of crystallized cloud and huge blocks of icy blue sky began to drop on everybody from above, and everybody was killed, the laughing rooster and the little red hen and everybody else in the barnyard, for the heavens actually *were* falling down.

—Moral: *It wouldn't surprise me a bit if they did.*

Preface

This book grew out of sense of mixed vexation and concern that has been building up inside me for a long time.

As a computer scientist I have been engaged for over thirty-five years in the modeling and computer simulation of systems and phenomena of all kinds. Over the years, I have worked on, taught about, and written about computer techniques to predict the behavior of air pollutants, space and aircraft, electric power systems, computer networks, the human heart, and managerial structures, among others. Many of the projects with which I was associated were successful in providing decision makers, managers, and engineers with useful information. Others turned out to be too vague in their formulation or too difficult or too expensive computationally, but all helped to teach me what can and cannot be expected from computer models. Moreover, I learned what we can now confidently predict and what we can hope to predict in the future.

I like bookstores and, whenever I get the chance, I enjoy thumbing through new books dealing with the future—predictions of what is to come and warnings of what might come. In most of my favorite bookstores, I have noticed several racks of books dealing with prophecies—books by "new thought" visionaries, by astrologers, and by practitioners of the occult. Many of these forecast dire events, such as major earthquakes, plagues, invasions from outer space, collisions with comets, and, not infrequently, the end of the world. I have paid very little attention to these.

Instead I have spent a good deal of time leafing through and occasionally buying books describing the future of our planet from the scientific point of view—books written by physicists, by chemists, by biologists, and by social scientists. These include books in sections devoted to the environment, health, economics, and sociology. A surprising number of books, year after year, feature forecasts of catastrophic events that will or might occur, along with exhortations to action to prevent or to mitigate the predicted calamities. These predictions came from my side of the house, from mainstream scientists working in major university or government laboratories. They were based on what is known as the scientific method: the systematic gathering of data, the formulation of a theory or hypothesis, and the validation of that hypothesis.

The subject matter of these scientific "gloom-and-doom" books changed from year to year. For many years, until quite recently, the threat and consequences of nuclear war took center stage; there was the energy crisis and the threat of running out of oil; there were urgent warnings of earthquakes and volcanic eruptions; there was the threat posed by agricultural chemicals; and almost every year there have been predictions of an imminent "repeat of 1929" and the collapse of our economic system. Some of these were clearly sensationalistic tracts, attempts to cash in on the public's fear and hysteria; others were oversimplified and superficial polemics. But every year a considerable number have been thoughtful and credible presentations by respected scientists intended to communicate the views of the scientific establishment to the general public. Written in a popular vein, they have skillfully mixed solid science with predictions of catastrophes that, while often overdramatic, have perhaps been justified to rally public support for an important and worthwhile cause. And I have been generally in support of most of these causes.

I have gradually become convinced, however, that these scientific predictions of imminent catastrophes have been doing more harm than good. To the casual reader, even to the involved reader, a "prediction" is a description of something that is going to happen or is at least likely to happen. But that is really not how scientists use the term when "predicting" catastrophes. Because they are based, without exception, on the extrapolation of theories and models into domains for which their validity has not been established, these predictions represent "scenarios" of what might happen if a great many assumptions turn out to be correct. It is difficult enough, even with the most powerful of computers, to predict the course of a natural phenomenon that is well understood and that is quite similar to something that has already been observed. But the catastrophes under discussion have rarely if ever occurred. So catastrophe predictions do not constitute "predictions" in the popular use of the term—yet they are interpreted to mean just that by the public. They therefore constitute, albeit unintentionally, warnings that mislead the public. And they may have some very damaging consequences.

There exists continuing competition among catastrophes. A new threat surfaces every couple of years, while one that has been receiving attention gets placed on the back burner and neglected before any meaningful progress has been made to counter it. Catastrophe threats compete not only with each other but also with the many other threats to society that have not yet been elevated to "catastrophe" status. They thereby effect a reordering of social priorities in what is often an illogical, unwise manner. Finally, when the threats don't pan out, the credibility of all scientific predictions, not just of catastrophes, becomes undermined.

It seems to me that by now the public—the well-intentioned and well-read part of the public—has been numbed by catastrophe predictions. It is as if you were told after your annual physical: "You have lung cancer, advanced arteriosclerosis, severe blood poisoning, and several dread infectious diseases. You better do something about these ailments at once, or else. And by the way, don't forget to pay the bill, though you are broke."

For many years I have kept my opinions to myself because I was afraid that they would be misinterpreted as criticisms of the causes that gave rise to the catastrophe theories or to suggest that they should receive less support. That is not what I believe.

But, by 1990, I had had enough. Starting on January 2 of that year and for the following eighteen months, I purchased every book featuring scientific predictions of catastrophes that I could find on the bookshelves of the large, general bookstores in my neighborhood, all books that seemed to be shaping public opinion during that time: a snapshot, if you will, of what was terrifying us most in the early 1990s. I tried to read all of them, but I concentrated on those that dealt with catastrophes that are likely to affect people all over the world, those that are global in character. There were close to 100 of these books in all, dealing with eight imminent catastrophes: the depletion of the ozone layer, the greenhouse effect, acid rain, nuclear radiation, the AIDS epidemic, overpopulation, earthquakes, and economic collapse. This minisurvey confirmed my conviction that catastrophe predictions should be treated much more cautiously by scientists and by the public. And so, this topic yields yet another book.

This book is divided into three parts, containing four, eight, and two chapters, respectively. In Chapter 1, I try to define what I mean by "catastrophe" and describe some of the mechanisms by which scientific predictions of catastrophes get started and catch the eye of the media and of the public. Chapter 2 is a digression— a brief survey of the unscientific methods of forecasting calamitous events that have been handed down through the ages, many of which are still in wide use today. In Chapter 3, the scientific approach to prediction is described, including a detailed discussion of the kinds of "models" that are used to that end. The "solution" of these models, the generation of predictions, usually with the aid of computers, is taken up in Chapter 4.

In Part II, the eight global catastrophes predicted in the books that were on sale in 1990 are examined one at a time. In each case I have based the treatment on one or two of the most authoritative and comprehensive recent books on the subject. I have tried to summarize the points of view of the authors in a concise and fair fashion. I do not suggest that the threats do not exist, that they are not very serious, and that they do not deserve substantial attention. I realize, of course, that books are not the only medium for the dissemination of scientific ideas. Catastrophes are staple fare for many magazines, movies, and television programs. But I believe that books, particularly the kind of books that I have selected, are the key element of the communication channel linking scientists with the most intelligent and influential public sectors. They form the basis for most of the other communications.

Part III includes some general conclusions and recommendations.

The reader of this book is assumed to be a scientifically aware individual interested in the environment, public health, and the economy, but he or she needs no special training in these areas. No background in mathematics or computers, for instance, is required. This book is primarily directed at the readership of the books that form the basis of Part II, and I have tried to treat very serious subjects in a sober way. Still, some readers may object to the fact that on occasion, on rare occasions, I may appear to deal with some rather calamitous subjects in a lighthearted manner. But when confronted by eight imminent catastrophes that promise to irradiate me, to fry me, to freeze me, to poison me, to pulverize me, and to impoverish me in short order, how can I greet tomorrow without an occasional touch of humor?

Acknowledgments

Many, many colleagues, students, and friends have provided me with invaluable information and criticism. I thank them all. I am particularly indebted to George Bekey, Andrew Charwat, Halina Charwat, Curtis Karplus, Mark Karplus, Lillian Larijani, June Myers, Robert Tisdale, Robert Uzgalis, Arlene Winn, and Ira Winn for reading large parts of the manuscript and for suggesting important improvements. Finally, I would like to express my appreciation to Linda Greenspan Regan of Plenum Press, who edited the book with exceptional insight, care, and patience.

Contents

PART I. PREDICTING CATASTROPHES

PART II. EIGHT IMMINENT CATASTROPHES

PART III. WHAT DOES IT ALL MEAN?

The Heavens Are Falling

The Scientific Prediction of Catastrophes in Our Time

PART I

Predicting Catastrophes

> . . . rising from the orchestra pit is the
> weather signal. . . . From time to time black
> discs are hung on it to indicate the storm or
> hurricane warnings.
>
> One of those black discs means bad weather;
> two means storms; three means hurricane;
> and four means the end of the world.
>
> —FROM THORNTON WILDER, THE SKIN OF
> OUR TEETH, 1942

ONE

What Are Catastrophes?

THE AGE OF INFORMATION STRESS

The rise of computers in the years following World War II is regarded by many observers as having initiated an "information revolution" rivaling in importance the industrial revolution of the nineteenth century. Just as the machines developed during the last century served to relieve humanity of arduous manual tasks, so the computers of the twentieth century act to unburden humankind of mental drudgery. In the very first stages of this revolution, computers took over many of the routine tasks formerly carried out by secretaries, bookkeepers, and the like. Gradually, however, computers have moved into jobs that were heretofore impossible to do because they were too time consuming. This is particularly true when it comes to scientific computing: scientists making computations to analyze systems and trying to predict what will happen in the future. A single modern computer can perform billions of calculations per second. Even if every man, woman, and child in the United States were given an adding machine and hired to do arithmetic, they could not achieve that kind of speed. So a new world has been opened to scientists.

One important consequence of the information revolution has been a very strong tendency in all scientific fields to become more quantitative in their approach. Formulas, numbers, and graphs have become the lingua franca of the scientist—in doing his* work, in communicating with other scientists, and in communicating with the public. This is true in the physical sciences such as physics, chemistry, and geology; it is true in the life sciences, in biology, physiology, and medicine; and it is also true in the social sciences including economics, sociology, and psychology.

In order to get a message across to the government or to the public, the modern scientist goes to the computer and generates streams of numbers, reams of graphs, and often colorful, animated graphics. The audience has been educated to expect this and is readily impressed and convinced. This shift from vague, qualitative generalizations to "hard" data and facts has worked out very well in most instances. Decision makers and the public that they serve are provided with a better understanding, a superior insight to permit them to better grapple with significant issues. But the pervasive computer has also created a new problem. So much data, so many "facts" can be generated with the push of a button that every one of us, every day is exposed to an overwhelming stream of quantitative information—economic data, election predictions, weather forecasts, advertising claims, medical studies, etc. Bombarded by numbers and graphs, we try to absorb and understand what we can, and sooner or later we give in to exhaustion and tune out. We live in an age of "information stress."

Scientists who feel that they have a significant message, one that must be gotten across to the public, are in a very difficult position. They are in a competition to catch the public's eye and the attention of the government. They are in competition with scientists championing other causes, and they are in competition with all the other promoters of products and ideas that flood the media with their data. One result is escalation.

---

*To avoid the awkward construction "his/her," I'll be using "his" for simplicity's sake.

Thus, the causes that get noticed, that get a hearing tend to be dramatic. It is not enough that the message be significant; it must be sensational. It is not enough to warn of an imminent danger; the warning must include a prediction of something calamitous, of something so terrifying that it takes precedence over the competition. At any point in time, society is confronted with a myriad of dangers and threats, and most of these threats fall into the bailiwick of some scientists or community of scientists that turn the struggle against that threat into a cause, their cause, *the cause*. Those that get attention have often made predictions of the most scary kind, predictions of catastrophes to come in the near future.

In order to gain a perspective of the information stress imposed on the public by catastrophe predictions—those made by eminent scientists—it suffices to scan the shelves of our neighborhood bookstores. Here are the main titles of some books that competed for the attention of the public during only one year, 1990 (full titles and authors [1–20] can be found in the Reference section near the end of the book and also in the Bibliography): *The Great Depression of 1990, The Toxic Cloud, The Silent Spring, The Population Explosion, Extinction, Global Alert, Final Warning, Deadly Deceit, Poisoners of the Seas, The Hole in the Sky, Hothouse Earth, The Great Dying, Ice Time, The Greenhouse Trap, Ozone Crisis, The AIDS Epidemic, The End of Nature, The End,* and many more. Each of these carries a crucial message and warnings of catastrophes to come, and each contributes to our information stress.

CATASTROPHES: DEFINITION AND SCOPE

In this book we are concerned with the use of models and computers to forecast very, very bad things. Not just run-of-the-mill bad things, but events of earthshaking consequence. What should we call these events? Every language contains many synonymous words or expressions. A common example concerns our northern neighbors, the Eskimos. Eskimos spend most of their time in a snowy environment; they live in snow igloos, they hunt in the snow, they are subject to snow storms—snow is very important to them. It is not surprising, therefore, that most Eskimo languages contain more than a dozen different words that all translate into the English "snow." There are words for dry snow, for wet snow, for old snow, for new snow, for dirty snow, for slippery snow, and so on.

In our Western civilization people have always been very interested in predictions of the future and especially in the prediction of misfortunes of all kinds. In early days, oracles, soothsayers, magicians, and astrologers were regularly employed to do this. The Bible and the scriptures of most religions are replete with prophecies of disasters. And now scientists are coming up every day with gloomy forecasts. Hence, most Indo-European languages have many different words to characterize the shade of severity that may be encountered, and English is no exception. Some of the more widely used of these terms are shown in the figure below. They have been placed in the order that current common usage assigns to them—going from bad to worse to worst. We will be concerned particularly with the shaded wedge labeled "catastrophic," the worst of the lot.

In Greek, the prefix *kata* variously signifies "down" or "over" or combinations of the two, and *strophe* indicates "turning." So *catastrophe* roughly means an "over-turning," a destructive revolution. In come disciplines, such as mathematics and geology, the term catastrophe has a much more narrow, specialized meaning. But we will use it in its general sense and recognize a catastrophe as an event or series of events having all of the following attributes:

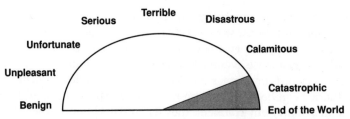

FIGURE 1. Adjectives describing the undesirability of events.

- Very *unfavorable* for a substantial portion of human society
- Extraordinarily *intense* in magnitude
- Exceedingly *rare* in occurrence
- Very *long-lived* in its effects on society

The anticipation of such a catastrophe elicits a number of typical responses from decision makers in governments and from the public. These usually include:

- *Panic* and other irrational behavior
- Frantic moves to *reorder priorities* in dealing with social problems
- Efforts to prevent or to mitigate the catastrophe "at any cost"

In this discussion we will use the term "catastrophe" to refer only to events that have a broad effect on *society at large*. From a personal point of view, a fatality is almost always a catastrophe of cosmic proportions, at least as far as the person who dies and his immediate family are concerned. A fatal airline accident is certainly a disaster for the passengers and also for the company. But by our definition, such an accident would be called catastrophic only if it were part of a series of accidents that somehow caused the demise of the public transportation system and greatly lowered the quality of our lives.

It is easy to look back into history and come up with events that are clearly encompassed by the above definition. In the ancient world, there was the Great Flood depicted in the Bible and the volcanic eruptions that caused end of the Minoan civilization and, in myth, of Atlantis. The Middle Ages had the great bubonic plague. In modern times there have been horrendous earthquakes and volcanic eruptions, the droughts and famines in the Sahel region of Africa, and the nuclear reactor accident at Chernobyl. Some of these events were more or less uniformly unfortunate for most of the people alive at the time. In others, it is a matter of the point of view. A war is almost always catastrophic for those on the losing side, but not for the winners. An earthquake may do terrible damage in one geographic area, but that may be good for business elsewhere. So in some cases, whether an event is catastrophic or not is in the eyes of the beholder. But for our purposes, in examining the great potential catastrophes of our day, we will generally limit ourselves to catastrophes that are *global* in character, that threaten to have a very bad effect on most, if not all, of the people of the world.

We will restrict ourselves in one more way. We will limit our discussion to catastrophes that are *imminent*, that are expected relatively soon. "Soon" in this context means within a few years or a few decades at most. Scientists are agreed that our world will come to an end in a few billion years, when the sun will expand and absorb all the planets. The approach of that end will certainly be disastrous for all living things at that time, but that does not make it an "imminent" catastrophe. On the other hand, if a scientist were to predict that Earth were to be swallowed by a black hole within the next fifty years, we would regard that prediction as being "catastrophic."

The eight catastrophes, all predicted by eminent scientists,* that will be the focus of our attention are:

1. Depletion of the ozone layer and the consequent increase of our exposure to ultraviolet light radiation
2. Change of the global climate due to increases in atmospheric carbon dioxide and other "greenhouse" gases
3. Pollution of the atmosphere through the emission of industrial chemicals
4. Pollution of the atmosphere by radioactive particles
5. The AIDS epidemic
6. Overpopulation and the population explosion
7. Worldwide economic depression
8. Earthquakes of catastrophic proportions

---

*In this book, the term scientist will be taken to include not only physical scientists, life scientists, and social scientists, but also computer scientists, applied mathematicians, physicians, engineers, technicians, and even the managers engaged in a scientific enterprise.

These eight catastrophes are not randomly chosen examples. They are *all* of the imminent catastrophes predicted by the scientific establishment in the early 1990s that attracted wide public notice. At least *I* was unable to unearth any others in the course of a far-reaching search of the popular literature in the English language. News of these predictions was not restricted to esoteric scientific journals or specialized meetings and conferences. All were presented to the general public in 1990 and 1991 in books written in a popular vein as well as in a large number of television and radio programs. And all have attracted the attention of government agencies and congressional committees.

THE NURTURING OF A CATASTROPHE THEORY

Scientifically based catastrophe theories are usually hatched in an academic setting or in the back rooms of government laboratories. An individual or a small group of scientists makes a startling observation or obtains surprising results from an experiment or from a computer run. This leads to speculations and inferences that suggest the possibility of a major disaster some time in the not-too-distant future. Still, most theories die early. Somebody, perhaps the scientist himself, discovers a fatal flaw in his logic or in the experiment. Or perhaps the scientist is too busy or too diffident to push his idea with sufficient aggression. But once in a while, when he finds upon reexamination no discernible flaws in his logic, he will take the next step and present his hypothesis to members of the scientific community at large.

New scientific theories are usually floated as trial balloons at meetings and conferences of scientific societies, perhaps at a local meeting or symposium, perhaps at a national or international gathering. Often there then occurs an outpouring of hostile or negative comment, enough to discourage the scientist from proceeding further. If not, the next step is publication in a scholarly journal and application for financial support from a government agency. Here the scientist is subjected to what is generally known as a "peer review." A journal editor or a manger in the government selects a handful of well-known and reputable scientists to comment on the paper or proposal and to judge whether it should be accepted or rejected. Of course there are many more rejections than acceptances. The scientist is free to submit a rejected paper or proposal elsewhere and to try again and again, but few have the persistence or the ego to keep trying. So the peer review process acts as an effective filter.

If a government agency, such as the National Science Foundation, the National Institutes of Health, or one of the departments of the federal government accepts a proposal, the theory is on its way. Money materializes to buy equipment or computer time, to hire research assistants, to do any necessary traveling—all to place the theory on a firmer foundation, to prove or substantiate assumptions and inferences, to obtain promising results, and to submit proposals for additional financial support. At some point, the scientist may find his theory proved wrong, in which case he abandons it to pursue other projects.

For theories that seem to be substantiated, the scientist, over time, may be awarded tenure or be rewarded with promotions and may achieve some renown in a specialized field. The idea or theory may be picked up by other scientists and institutions, all competing for money, and eventually many scientists may participate in exploring the subject. But only on rare occasions does a theory break through to the next level of importance or notoriety—recognition by the world at large.

Now the scientist is in a new, daunting arena. Reporters come to call, and articles begin to flourish in newspapers and magazines. Established companies and venture capitalists may become interested in practical applications. Activists of all sorts may rise in support of or in opposition to the theory, and eventually politicians may get into the act. This may be an exhilarating time for the scientist but also a period during which he eventually loses control and becomes a spectator, a participant, or perhaps a figurehead of a movement that seems to have a life of its own. An advocacy community is born.

THE RISE OF ADVOCACY COMMUNITIES

What is an advocacy community? In addition to its most common connotation (all people living in the same geographical area), the term "community" has a number of different shades of meaning. We speak of the "scientific community," the "diplomatic community," the "intelligence community," the "academic community," etc. Here the word is used to identify people engaged in the same profession, usually a lifetime commitment. Not so for "advocacy communities."

By contrast, an *advocacy community* is an informal but coherent grouping of individuals united by a common cause and objective. Many such communities are based on religious beliefs, others on ethnicity, and still others where the members share common social or political principles. In this book we are concerned with advocacy communities of a special kind: communities that are based on a scientific theory and that advocate causes that are inspired by that theory and its implications. I will call such science-based advocacy communities "Communities," for short.

"Members" of the Community include the scientists actively engaged in the exploration of the theory as well as those studying its implications. But membership also includes the people in government agencies who are directly involved in the granting of the funds necessary to carry on the work. In addition, it includes scientists and engineers in companies that hope to capitalize on the idea in one way or another. Then there are consultants and consulting companies who sell or who hope to sell their advice on this theory to the government and industry as well as the individuals and groups within national laboratories and think tanks whose job it is to monitor and to evaluate the theory as it evolves. So the scientists who gave birth to the theory and who were central to its initial development eventually comprise only a minute and often uninfluential part of the Community. At its peak, a Community may have a membership in the thousands and may well become a formidable force in academia, industry, and the government. If the theory has earthshaking implications, such as the prediction of an imminent catastrophe, the Community usually acquires a large, devoted, and assertive following among the public at large—active supporters numbering in the hundreds of thousands or more. Unlike so-called vested-interest groups that are formed primarily to protect and enhance the wealth and power of its members, most Communities, including those associated with the eight catastrophes discussed in this book, are largely motivated by unselfish, humanitarian considerations. And unlike vested-interest groups, Communities are ephemeral. A Community is formed when the need arises, and it flourishes as long as the theory that gave it birth shows promise and receives support. When the theory is "shot down" or when support and public interest wanes, the Community shrivels and may die—many of its members quickly shifting their attention elsewhere, perhaps to other Communities. An active scientist is not infrequently a member of several Communities at the same time, and a single government agency may well house hundreds of such Communities.

An example: In my own long and diversified career, I have been a member of a dozen or so Communities. In most I participated as an active researcher, in a couple as an investor in manufacturing and service companies, and still in others as a technical consultant. During the 1980s I served for several years as a member of a committee formed by the National Research Council, an arm of the National Academies of Sciences and Engineering. The charge of that committee was to evaluate and to report on issues related to electromagnetic pulse effects, the EMP threat. Much of the work was hush-hush, but a report[21] was eventually published and released to the public. Here is the situation.

When a nuclear bomb is detonated at or near ground level, the explosion generates a lot of heat and radioactive fallout, as well as the notorious mushroom cloud. Not so when a nuclear device is set off in outer space [at elevations higher than say fifty kilometers (thirty miles) above sea level]. Then there is no noticeable increase in temperature or in radioactivity at ground level. Instead there is a highly intense, very short-lived electric current—a lightning bolt. That bolt, known as an *electromagnetic pulse*, *EMP* for short, lasts only a small fraction of a millionth of a second. That is too short to harm living things, but the EMP is so powerful that it can do an enormous amount of material damage.

The theory is that a single well-designed nuclear device detonated thirty miles above Nebraska could knock out the electric power grids throughout the entire continental United States, immobilize all aircraft and automobiles, disable most computers and electronic apparatus, disrupt all telephones, radios, and other forms of electronic communications, and that's just the beginning. It would be a nightmare for the public in general, but particularly for the military that has long cherished the concept of deterrence. How can the president order appropriate retaliation without a working communications network? How can the country react to an attack if all telephones, radios, airplanes, and missiles are made inoperative? Hence, there is great concern about EMP and how to "harden" electrical equipment so as to prevent its effects. By the 1980s annual expenditures for protection against EMP amounted to billions of dollars.

Our committee met regularly and received detailed briefings regarding EMP and how to protect against it from many scientific, industrial, and military specialists. In their presentations the speakers referred frequently to "the community": "This or that is in the best interests of the community," "The community is convinced that. . . ," and so forth, using the term "Community" to include all those who had a stake in assuring that the study of EMP and the hardening of equipment received sufficient government support. There was also occasional mention of "enemies of the community"—not Russians, but American physicists who cast doubt on the existence of EMP or on the supposed severity of its effects. And in the course of time, all of us on the committee came to understand that we too had become members of the EMP Community.

Late one afternoon in Washington, as speaker after speaker droned on, my attention began to wander and I had a vision. I saw in my mind's eye a conference room in Moscow in which a panel of aging civilians was subjected to a presentation on the subject of EMP. I even saw a professor, just like myself, tired and bored, near the back of the room, hearing the same kind of material as I was hearing. And it occurred to me that we were all members of the same Community. Certainly we were patriotically concerned with the safety of our respective homelands. But above and beyond all that, we were all committed to persuading our governments to provide adequate or more than adequate funds for EMP. And for that we needed each other.

In the end, though, our eight-person committee buckled down and ground out a report that, while not making big waves, did constitute a fair and evenhanded assessment of the EMP problem. It represented a balance of views from inside and outside the EMP community and provided insights on the following questions:

- How real is the EMP threat?
- How credible are available computer models and simulations?
- How effective are alternative "hardening" methods?
- How sound are the current programs to "harden" military equipment?
- What new research and development efforts should be undertaken?

These are questions of vital significance to decision makers at all governmental levels, and their answers are determining considerations when it comes to deciding which government programs should be nurtured and which should be cut or eliminated. At any given time, dozens of such committees are similarly engaged in illuminating scientific and technological issues and their reports fill many Washington bookshelves. These reports have been very influential in shaping the opinion of government decision makers.

The Positive Side. It would be a big mistake to infer from the above anecdote, that Communities are principally self-serving boosters of causes that are in their own interest. There is a small degree of that to be sure, but Communities play a vital, indispensable part in the development of a theory, in its reduction to practice, and in its practical implementations. In brief, a community helps to establish and to maintain order. Here are some of the important roles that it plays.

When a new and promising theory emerges from its incubation period, when it becomes a Theory, it achieves a measure of respect from the world outside a narrow scientific specialty. If the Theory later becomes "hot," there is often a rush to climb on the bandwagon. That wagon becomes very crowded indeed when the government begins to allocate substantial funds to the elaboration and development of the Theory. The competition for fame and fortune is joined not only by academics looking for research support and by industry types looking for lucrative contracts, but also by individuals and groups within the government. There results a veritable flood of proposals for new theoretical and experimental investigations, for new gadgets, for new laws and regulations, in short for everything under the sun. And all these proposals have to be sorted out.

In the beginning, the staff of the funding agency and the editors of journals can cope, but gradually they become overwhelmed. The task of filtering, at least the preliminary filtering of ideas and proposals, then devolves on an emerging Community. Often the scientists most closely involved with the Theory convene small symposia or conferences to articulate their ideas and to let others be heard. Reports, technical papers, and eventually books are published and circulated. After a while others from academia and industry join in, and the Community grows and engages in all sorts of activities, but one of its principal motivations is to guard and protect the Theory. Though, if the Theory turns out to be wrong, the community will drop it as quickly as you can say "cold fusion."

Very quickly, informally to be sure, the Community adopts a set of standards and rules of practice: what historical figures should be respected, what instruments and measuring techniques can be trusted, what models and computer simulations are appropriate, etc. And all members of the Community are expected to conform. And they do. Members of the Community may work for competing companies, they may be rivals for tenure slots at a university, they may hate and fear one another, but to the outside world they present a dignified front. They are the statesmen to be consulted on all matters pertaining to the theory.

One of the functions of a Community is to specify the proper way of presenting and of packaging a new suggestion or proposal. This serves as an informal guideline to individuals wishing to enter the Community or to contribute to its work. And it provides a yardstick to evaluate new ideas. On occasion, however, a worthwhile idea may be rejected or may not even get a hearing because its author is not a member of the club or does not conform. But by and large the system works, and appears to work much better than any practical alternative.

Hence, Communities are essential. If they did not exist, somebody would have to invent them. They serve as an interface and a buffer between the Theory and the rest of the world. However, we must not lose sight of an important point. When a person becomes associated with a Community, he or she joins a team and becomes committed to its objectives and ideals. This is generally for the good of the Community's cause. To be truly creative and productive a person needs to be fervid and enthusiastic, the more so the better. But he must take great care not to lose too much of his objectivity and global perspective, which is not always easy.

The members of a Community thus become partisans of a cause, *the* cause, often a controversial cause. We would probably not expect the quarterback of a college football team to opine sagely and dispassionately on the pros and cons of intercollegiate athletics. Similarly, members of a Community are perhaps not the best people to ask to compare the importance of their Theory to theories espoused by other Communities. They are expected to be quarterbacks and cheerleaders at the same time. In that spirit, some of the members of a Community, bubbling with commitment, may make catastrophic predictions, the nuances of which are lost on the public. And in making these predictions, they touch a sensitive nerve throughout the general public.

WHY CATASTROPHE PREDICTIONS CREATE A STIR

Predictions, especially predictions of catastrophes, have always occupied a very important place in the psyche of the human race. Ancients and moderns alike have been fascinated by prophecies, not only out of curiosity about the future, nor primarily to learn useful information. In fact, from the days of the Delphic oracles, to the Roman vestal virgins, to the Revelation of St. John, to Nostradamus, prophecies were invariably so convoluted, ambiguous, and murky that few could interpret them until after they were fulfilled—they were always fulfilled. Rather, prophecies were valued because they confirmed the existence of order in the universe. Fate is in the hands of the gods, and all important events in the future have been decided once and for all, chiseled in stone. A mere mortal, even a tragic hero such as Oedipus, might struggle bravely, but in the end he was sure to be defeated by inexorable fate. That view certainly held people in awe, but it also provided a measure of comfort. Rare is the person who can read the following lines without experiencing a mixture or ambivalence of feelings:

> DEATH: There was a merchant in Bagdad who sent his servant to the market to buy provisions and in a little while the servant came back, white and trembling, and said, Master, just now when I was in the market-place I was jostled by a woman in the crowd and when I turned I saw it was death that jostled me. She looked at me and made a threatening gesture; now, lend me your horse, and I will ride away from this city and avoid my fate. I will go to Samarra and there death will not find me. The merchant lent him his horse, and the servant mounted it, and he dug his spurs in its flanks and as fast as the horse could gallop he went. Then the merchant went down to the market-place and he saw me standing in the crowd and he came to me and said, Why did you make a threatening gesture to my servant when you saw him this morning? That was not a threatening gesture, I said, it was only a start of surprise. I was astonished to see him in Bagdad, for I had an appointment with him to-night in Samarra.
> —From W. Somerset Maugham, *Sheppey*, 1933
> (Used by permission of A.P. Watt Ltd
> on behalf of The Royal Literary Fund.)

On one hand there is the dreadful finality of it all; on the other hand there is a measure of relief. If your name is not on Death's current list, you are safe at least for the present. Be pious, trust and submit, and lead your simple life in peace. That is one of the morals of the anecdote.

Most of the ancient and classical prophecies, at least those that we know of, forecast misfortune. That is to be expected. If a hero is to be pictured as struggling against fate, it makes sense that fate has some calamity in mind for him; otherwise he would not be motivated to struggle. In the Christian era, and particularly after the Reformation, additional elements took on significance: guilt and punishment. The post-classical tragic hero exercises free will, transgresses, and sins. The prophesied tragedy is just punishment and demonstrates that not only order but justice as well reigns in the universe. The prophesy is a confirmation of religious and moral principles. That appears to be an important ingredient of present-day prophecies in the United States and Europe.

Americans are known as an optimistic people. The Horatio Alger story, the notion of Yankee ingenuity, the philosophy of rugged individualism—all contribute to the prevailing myth that all is possible with sufficient imagination and hard work. The shelves of our bookstores are replete with self-improvement and "how-to" books. Positive thinking is at the top of most prescriptions for improved health and wealth and contentment lists. Yet there is also a persistent countercurrent, a fascination with the dark side.

Many devout people from diverse religious backgrounds believe in biblical prophesy. A significant number of them are convinced that the Book of Revelation provides the key to very detailed forecasts of the immediate future. Studies show that between 1800 and 1960 well over one hundred religious sects and cults were formed specifically to prepare for an imminent end of the world—always within twenty years or less. Obviously the prophecies that motivated these cults all turned out to be incorrect; they remained unfulfilled. But these failures did not appear to shake the faith of the prophet or that of the disciples. New prophecies are made and new countdowns commence. Not only within these cults, but also in the larger, more mainstream religious environment, calamities such as earthquakes, plagues, even UFO sightings frequently have been and will continue to be seen as handwriting on the wall—as just retributions for the sins of humankind. As the third millennium approaches, watch for a proliferation of such dire forecasts.

Most scientists stay clear of religious prophets and prophecies, at least while at work, as do most government decision makers. Scientists rely on the scientific method and computer models to make their predictions, and their predictions are rarely precise and clean-cut. Often scientific predictions are based on so-called *scenarios*. The scientist makes a set of assumptions, programs a computer using these assumptions, and the output predicts how the future will look, but only if the assumptions are correct. Because the scientist is usually uncertain about the assumptions, he will make many computer runs, each making different predictions, because each is based on different assumptions.

For example, in simulating the global climate, it is necessary to guess or to assume how much carbon dioxide will be released into the atmosphere over the next several years. That depends heavily on unknown factors such as future energy consumption, the introduction of alternative energy sources, the future of nuclear power, etc. The more carbon dioxide that is assumed to be released into the atmosphere, the greater the undesirable changes in the climate. Many scenarios are run on the computer and many different forecasts are made—some predicting small changes and others dramatic changes. In this particular example, all of the scenarios yet implemented on computers predict bad news to a greater or lesser extent.

Scientific forecasts are almost never based on a single forecast of the future, but on many (perhaps hundreds or thousands) "predictions" in the form of stacks of computer outputs. The simulation results are very valuable to the scientist. They help him to understand better the phenomenon being modeled and also to get an idea of the possible range of undesired effects, such as possible increases in the average global temperature. No single computer run or computer output can be considered to be a definitive forecast of what will happen, although some "predictions" may be regarded as more probable than others. A few of the "predictions" may be very ominous indeed, even catastrophic.

When a Theory leads to "predictions" that suggest an imminent disaster, the scientists and the Community to which they belong are unlikely to keep the bad news secret. Usually, the media are somehow alerted and are not averse to serving as the bearers of spectacularly evil tidings. Popular books soon appear and some become best-sellers; ultimately the executive and legislative arms of the government are compelled to take notice. In most instances the scientists and their Communities are motivated first and foremost by humanitarian considerations. The public should be warned, and the government should promptly take all possible steps to prevent or at least to alleviate the effects of the predicted calamity. Unfortunately, the scientific bases and the real significance of the publicized predictions often get lost in the shuffle.

It appears that the public is generally complacent when presented with pessimistic (not quite catastrophic) forecasts. Incremental unfavorable changes in the climate, public health, or the economy arouse only a moderate concern. But when a Catastrophe is predicted, terror strikes home. The threshold beyond which the merely disastrous becomes catastrophic is not easy to discern, especially in advance; but it is there. It calls forth an atavistic response not unlike that of folks in days of old to prophecies of floods, plagues, famine, and war, and not unlike the response of many moderns to prophecies of the end of the world. There is the same dreadful foreboding and the same mixture of guilt and satisfaction that the guilty are going to receive their just punishment. Overreaction and even panic take over.

This mood is not only apparent when scientists forecast natural disasters, however. For well over forty years, most economists have warned that the perilous excesses in the United States' financial system—continual deficit spending, the burgeoning national debt, the rich getting richer, and the like—will have the direst of consequences. One well-established Community within the economics discipline adheres to a Theory that predicts the complete collapse of the American economy, of the world monetary system, and even of many of our cherished social and governmental institutions—all this to occur within the next two years. This catastrophe is deemed to be inevitable. All that an individual can do is protect himself and ride the storm, while the rest of the world disintegrates. This same theory has been trumpeted almost continuously for many decades, its vigor undiminished even as the onset of the apocalypse is postponed again and again.

I remember attending several daylong seminars sponsored by that particular Community. They featured lectures by eminent economists, libertarians, writers, and successful financiers. The common thread running through all of the presentations was that the unavoidable catastrophe would soon be upon us, and all listeners were counseled to hoard gold, Swiss francs, food, water, and bullets to survive the coming dark days. Most of the audience was enthralled, ready to attend follow-up seminars and new programs, apparently not chagrined by the failure of earlier predictions based on the Theory. Certainly some listeners, including myself, were impressed by and admired the economic principles, the logic, and common sense on which the Theory was based. Though a large segment of the audience believed that—we were living beyond our means, we were enjoying an undeserved prosperity, we had traded spirituality for materialism—we were all sinners and that we would all be punished. *All*, that is, except for the initiated few who would ride out the storm in a modern equivalent of Noah's ark.

WHERE IS THE HARM?

The communities that back the eight catastrophes that are the focus of our attention in this book, all and without exception, champion laudably worthy causes and ideals: a cleaner and less hostile environment, improved health, a prosperous economy, and so forth. Hence, the following thoughts may well come to mind.

Suppose that the prediction of a catastrophe is off base, and the heroic efforts made to prevent or to mitigate the catastrophe are not necessary. Is it not also true that these efforts are nonetheless exerted in the right direction and achieve worthwhile results? We all know, for example, that smoking is unhealthy. Is it such a bad idea then to publish antismoking ads that graphically show the dangers of smoking and scare people? This can only save lives and it hurts no one, except those in the tobacco industry. The more people that are restrained from smoking, by whatever means, the better. Similarly, suppose that it turns out in a few years that catastrophic global warming due to the greenhouse effect was really not in the cards at all. Surely, if the prediction of a global warming catastrophe encourages governments and industries to get together and to reduce the emission of carbon dioxide and other greenhouse gases—well then, great. Even without a catastrophe, the buildup of carbon dioxide in the atmosphere is very undesirable, and reasonable means used to reduce it should be applauded. So the argument goes.

We will have more to say about the possible negative and destructive aspects of catastrophe predictions in Part III of this book. Here is some food for thought in the meantime.

By definition, an imminent catastrophe poses a great threat to society, so great that it demands and often causes a reordering of social priorities. The fight against the catastrophe is moved to the head of the line and everything else has to wait its turn. In that sense, the prediction of a catastrophe can force the hand of government decision makers.

For example, a catastrophic earthquake has been predicted for California: "The big one is coming." To mitigate the potential destruction and loss of life, crash programs have been initiated to strengthen and modify buildings and other vulnerable structures all over the state. Of itself, this is surely a worthy effort and one that few will regret, even if the "big one" does not come within the next fifty years. But what about other worthwhile and vital projects that have to take a backseat while available funds are spent on earthquake safety? In the climate created by the earthquake catastrophe prediction, it takes a very rare and courageous administrator to channel funds away from building safety and into, say, low-cost housing for the homeless—even if he firmly believes that the suffering inflicted by inadequate housing outweighs the potential grief due to an earthquake. Administrators simply cannot afford the risk of being accused of negligence if the earthquake does come; they swallow their reservations and choose the politically astute course.

Consider the AIDS epidemic. When celebrities stage fundraisers to benefit AIDS research, the money that is contributed by their fans would most likely otherwise have been spent for other entertainment purposes; so that is all for the good of society. AIDS is a horrendous disease that must be conquered. But when the AIDS Community exerts strong political pressure on the National Institutes of Health (NIH), the result may well be an increase in the funding of AIDS research at the expense of support to combat many other pernicious diseases such as childhood leukemia, diabetes, or multiple sclerosis. In essence, the directors of NIH are compelled to suspend their best and fairest judgment in order to respond to the fears of an AIDS catastrophe.

From a broader and less dramatic point of view, every time a Community raises storm signals and cries "catastrophe," and the prediction doesn't come true, it erodes the credibility of scientific predictions and of science in general. Over the past hundred years, the scientific method has been shown and recognized to be the most effective and reliable route to progress and to the attainment of goals in most areas of societal importance. Science has won its spurs by virtue of its successes and achievements. When events are predicted by scientists, the public and the government expect these predictions to be valid.

In recent years there have been a number of investigations of how the public perceives catastrophe predictions and how it responds when these predictions fail to materialize. A case in point is Ralph Turner's[22] *Waiting for Disaster*, in which he presents a detailed sociological analysis of an earthquake scare in California in the late 1970s. This and similar studies suggest that scientists enjoy a large reservoir of credibility. The public seems to have faith in science and in the predictions scientists make even when these predictions fail to pan out. But that faith surely has its limits. If scientists and their Communities cry "wolf" too often and incorrectly predict catastrophe after catastrophe, people are going to stop believing—stop believing not only in catastrophe predictions but also in predictions (and prescriptions based on the predictions) that have far less uncertainty attached to them. And that would be very sad indeed. Not catastrophic, but very unfortunate just the same.

TOWARD A MORE RATIONAL USE OF PREDICTIONS

In our modern, terrifyingly complex world, reliable predictions of future events have become indispensable. Decision makers in government depend on these forecasts in their day-to-day activities as well as for long-range planning. And so does the public. The methods used by scientists have, in general, proven to be far more effective and reliable than alternative unscientific ways to that end. It is only when scientists use their models and computer simulations to predict the extreme events that we call "catastrophes" that things go astray. Under these conditions, the implications of the predictions become easily misunderstood and often evoke irrational responses on the part of the government and the public. Such irrationality is a threat to the long-term welfare of our society, even when the scientific predictions of a catastrophe are well-intentioned and when the reactions that they produce appear to be for the best. For this reason, it is essential that all those who make use of scientific predictions, and that includes nearly all of us, have some insight into the ways that scientists make predictions. That is my motivation in writing this book.

In order to gain some perspective, in Chapter 2 we first take a brief look at the methods of foretelling the future that have been handed down to us through the ages. These are still in wide use today and have always attracted and will always attract a large following. Then, in Chapters 3 and 4, we will explore in some detail how physical scientists, life scientists, and social scientists use models and computers to make their predictions. These two chapters may be hard going in places, but they are really the meat of the book. In Chapters 5 through 12, we examine the eight catastrophes that many respected scientists now tell us are coming soon. The objective of that discussion is to critique neither the models used by the scientists nor the causes that they have inspired. All the causes are worthy and deserve support. Rather, we want to come to understand how the models used to make the predictions fit into what I call "the spectrum of models" and how that affects their credibility. In the final two chapters of this book, we will try to reach some more-general conclusions.

TWO

Unscientific Predictions

Let's begin our discussion of the prediction of catastrophes by taking a brief look at methods of prophesying that have been handed down through the ages, methods in vogue at the dawn of our civilization and still widely practiced today—religious prophecies, magic, astrology, and the like. One thing that they have in common is that they are not based on modern science. They are unscientific.

This book is about scientific predictions, the use of computer models and simulations, high tech. It is addressed primarily to readers who have been weaned on the scientific method and who look to scientists for advice and counsel: readers who consult meteorologists for weather forecasts, not the *Farmer's Almanac*; who go to physicians when they are ill, not to faith healers; who have trust in seismologists when it comes to earthquakes, rather than in astrologers; in short, readers who, while aware of the many limitations of modern science, nonetheless respect the scientific way of thinking rather than the alternative. Why then begin with some comments on unscientific predictions, predictions that are not based on what we regard as established science? Actually, there are three reasons.

The first reason is historical. From the beginning, people have asked searching questions: How can we deal with the forces of nature? How can we treat illnesses? How can we organize our governments and societies? How should we lead our lives? They usually turned to respected and gifted individuals for answers. Modern science is essentially a product of the period of enlightenment that followed the Dark Ages and the Middle Ages, before science, alchemy, magic, and pseudoscience prevailed. Around the fifteenth century, there was a gradual transition from the unscientific to the scientific, and early scientists dabbled in both. For example, the great Isaac Newton, whose work provides the cornerstones for many modern scientific fields, also worked on and wrote copiously about biblical prophecy. The scientific methods that we will discuss have historical antecedents.

The second reason is psychological. We all live in a world in which the scientific and the unscientific coexist. We are exposed daily to modern prophecies and prophets. Astrologers forecast earthquakes and other disasters, Nostradamus's quatrains warn of epidemics and revolution, and there is no dearth of prophecies forecasting the end of the world—soon. Many people believe these prophecies, and many more half believe or at least pay attention to them. The media give them widespread notoriety. A scientist using scientific techniques to forecast a catastrophe, therefore, usually finds himself in competition for the ear of the public—a public that has been conditioned to a continuing parade of doomsday prophets.

We will come to the third reason at the end of this chapter.

PROPHESYING THROUGH THE AGES

The term *prophet* connotes any individual who speaks for God or the gods, who serves as a mouthpiece for divinity. Prophesying, therefore, is not limited to forecasting what is to come. In fact, the Old Testament prophets talked more about the past and their present than about the future. But in this chapter, we are concerned specifically with the prediction of future events, and particularly with the use of the supernatural and the occult to that end.

Supernatural is defined as something existing outside the normal experience and knowledge of humans, something caused by other than the known forces of nature. In early and primitive societies, where there was little conceptual knowledge of nature and no formal science, the supernatural played a major role in most day-to-day activities. Every individual was somehow in contact with gods, spirits, demons, and the like, and communicated directly with them by conversation or by various magical devices. As societies became more sophisticated, they also became more specialized, and a few gifted people were somehow chosen and charged with the task of communicating with the supernatural on behalf of the rest. Not infrequently, these select few formed an elite that passed occult and esoteric knowledge from generation to generation. And these initiates served their kings, their leaders, and the people by predicting future events as well as by advising them on a wide spectrum of issues and problems.

In the modern era, science gradually illuminated many of the mysteries of nature, and scientists took over most of the functions formerly performed by the mystics. Yet, to the present day, religion, magic, and the occult continue to maintain a strong hold on governments in many parts of the world and on the public at large. In this section we take an uncritical look at some of the principal ways that the supernatural has been and continues to be invoked to provide prophecies of what will happen in the future, with emphasis on European and American thought.

Signs, Portents, and Magic. People in all known civilizations have attempted to invoke the supernatural to discern the future course of events. Eventually, organized religions provided the preferred avenue to that end. But early on, it was the sorcerer, the magician, or the witch doctor who was sought out by rulers and common folk alike. The term *divining* is used to characterize the prediction of future events using information obtained from signs, omens, dreams, visions, and various kinds of artifacts. Invariably it is a special "gift" or secret knowledge that enables the seer to make the correct interpretation of the clues. According to R. Guiley,[1] there are two types of divination.

One branch of the divining arts is based upon the interpretation of patterns observed in the natural environment. The editors of the Time Life series *Mysteries of the Unknown*[2] assembled a sampling of forty-four different means of foretelling the future that have been used over the centuries. These range from *aeromancy* (forecasting by observation of atmospheric phenomena) to *hippomancy* (by the gait of horses) to *phyllorhodomancy* (by the sounds of rose leaves) to *xylomancy* (by the appearance of fallen tree branches) and finally to *zoomancy* (by reports of imaginary animals).

The other form of divination involves the interpretation of patterns created by the seer. Sometimes sticks, stones, or grains are thrown onto a surface, and the arrangement that these objects assume is "read" to tell the future. In the Middle Ages, Tarot cards became a favored technique of fortune-telling. Often seers find inspiration by gazing at reflective surfaces such as still water or mirrors or into a crystal ball.

The Greek god Hermes is credited with the authorship of the forty-two books that form the basis of the most widely used occult practices of our day, at least in the West. The Hermetica was probably the work of a number of ancient authors, and most of the books were destroyed long ago. But many medieval and modern occult movements and practices trace their origin to the knowledge revealed by Hermes Trismegistus—actually a combination of the Greek god Hermes and the Egyptian deity Toth.

Messages from the Past. There is a persistent belief among those who are mystically inclined, that our civilization was preceded by an older, more-developed, wiser race. Perhaps that race lived on the lost continent of Atlantis, perhaps they were visitors from outer space. In any event, they left behind traces in the form of stone monuments all over the world: Stonehenge, the Egyptian pyramids, monoliths in India, elaborate structures and works of art in Central and South America, etc. And, so the myth goes, they had the gift of prophesy. They could see the future of our planet—to the very end. Unselfishly, these superior beings strove to transmit their knowledge by embedding it in the monuments that have survived to our day. It is only a matter of "breaking the code," learning to decipher the prophetic messages, and the future will become an open book.

This approach to prophecy is best exemplified by the Egyptian pyramid of Cheops, the Great Pyramid. That pyramid has fascinated mystics and philosophers through the ages. But not until the eighteenth century did a cult of pyramidologists become established in Western Europe. According to this cult's tenets, the Great Pyramid was built to transmit a divine revelation or prophecy to the modern world. A message of turbulent times and catastrophes to come with a precise indication of the dates to the precise day, month, and year of these events. This message is carried by the dimensions of the pyramid, its passages, alleyways, and chambers. According to the pyramidologists, everything in the pyramid is laden with meaning. By very careful measurements, and perhaps with just a touch of imagination, they are able to discern all of the major events of human history from biblical times to the present and into the future. The Exodus, the Great Flood, the Crucifixion, World War I, the Depression, World War II, and so on. And by now we are perilously close to the end of the corridor, to the end of history—*the* End.

Astrology. Humanity's earliest attempts to foretell the future by observing the movement of the planets against the background pattern of the stars are shrouded in antiquity. Stone monuments, such as Stonehenge in England, dating from 2500 B.C., and the Egyptian pyramids appear to have been designed at least in part to serve astrological purposes. Babylonians and Assyrians recorded astrological omens on stone tablets as early as the eighth century B.C. For the next several hundred years the Babylonians used astrology principally to predict major events such as wars and floods as well as their effects on the state.

By the third century B.C., astrology had been exported to Greece, where it was systematized and popularized—horoscopes for the masses. In the second century B.C., Claudius Ptolemy, one of the most prominent intellectuals of his time, formulated the principles of cosmic influence, the way the planets govern human affairs. He introduced the use of planets, houses, and the signs of the zodiac, and his scheme for astrological prediction has remained essentially unchanged to the present day. (For a readable survey of the subject, try *The Compleat Astrologer* by D. Parker and J. Parker.[3])

Astrology was widely used in the heydays of the Roman empire, then fell into disuse during the early Christian era, only to be revived at the end of the Middle Ages and during the Renaissance when it was patronized by popes and kings alike. The advent of rationalism and modern astronomy, championed by Kepler, Copernicus, and Newton in the seventeenth century, initiated a two-hundred-year decline in the fortunes of astrology, at least in the Western World. Astrology was resuscitated once more in the late 1800s and has enjoyed growing popularity since then. So-called "modern astrologers" avoid specific predictions; instead they content themselves with determining favorable and unfavorable periods of time for persons and nations, moments of great danger, and times of good fortune.

Voices of the Gods. When the rulers of ancient nations needed advice they visited oracular shrines and consulted the oracles. Such shrines were prominent in Mesopotamia, Babylon, Egypt, and throughout the Middle East. Each shrine had its distinct modus operandi. Some of the oracles divined by observing signs, such as the flight of birds or the movement of the leaves of sacred trees; and some interpreted dreams. As time went on, though, most oracles provided spoken messages from the gods, often while they were in some sort of a trance. Oracles attained the peak of their influence in classical Greece and later in the Roman empire. The famous oracles of Dodona and of Delphi were staffed by carefully selected young women who inhaled sulfurous volcanic fumes to heighten their receptivity. Tradition has it that the oracles were never wrong. The problem was that their advice and predictions were often so murky that the advisee could not decipher their true meaning until it was too late.

A host of prophets carried on the tradition of the oracles through the Middle Ages and into modern times. Historical records include the names of hundreds of individuals of many nationalities and many faiths who made wondrous prophecies of things to come. For example, the Book of Merlin, a twelfth-century compilation of Welsh prophecies, exerted a significant influence throughout Europe for several hundred years. But the most famous of all European oracles was Michel de Nostradame, more widely known as Nostradamus.

During his lifetime, this sixteenth-century physician composed almost one thousand four-line poems, called quatrains, each a prophecy of future events, mostly dealing with the fate of European nations and their rules. According to one modern interpreter,[4] because he wrote at the time of the Inquisition, "Nostradamus confused the dating and ranking of his predictions purposely, and wrote them in a truly bewildering mixture of symbols, Old French, anagrams, Latin and other literary devices, so as to be able to fool the investigators of the Inquisition . . . Nostradamus did not leave a 'key' to his predictions. [This] has led to some curious and widely varied versions of his quatrains. Yet many of them are so close to fact that his reputation as Europe's greatest psychic continues undiminished."

In the tradition of the Greek and Roman oracles, Nostradamus's predictions are so convoluted, vague, and imprecise that they can only be certified as "true" and fulfilled after the fact, when it is too late to take advantage of them. Each generation has had its enthusiastic Nostradamus interpreters who found that many of the predictions applied to them and their times, and many of the quatrains have been used over and over in this way.

To get a flavor of Nostradamia consider the following translation of one of the quatrains:

The assembly will go out from the castle of Franco,
The ambassador not satisfied will make a schism:

> Those of the Riviera will be involved,
> And they will deny the entry to the great gulf.

and its interpretation by Stewart Robb[5]: "By means of a correct reading of this extremely lucid prophecy I was able to forecast: that Generalissimo Franco would meet the Axis powers on the Riviera, that there would be a crucial meeting, and that Hitler and Mussolini would fail to get Gibraltar. . . . The odds against this prediction being a chance coincidence are very high." He also feels that the following quatrain "is so clear as to need no interpretation" in its prediction of the 1917 Russian revolution:

> Songs, chants and slogans of the Slavic people
> While princes and Lord are captive in the prisons,
> In the future, by idiots without heads
> Will be received as divine oracles.

Since the days of Nostradamus innumerable visionaries, prophets, and would-be prophets have caught the eyes and ears of the public, and a number of their prophecies have been fulfilled in all or in part.

Biblical Prophecies. In the Western World, the Bible has been by far the richest source of prophecy. In his *Encyclopedia of Biblical Prophecy*, J. Barton Payne[6] catalogs 1239 predictions in the Old Testament and 1817 predictions in the New Testament. These have been the subject of countless interpretations and reinterpretations over the past three thousand years and have given rise to thousands of religious cults and movements.

The last book of the New Testament, the Book of Revelation, is totally devoted to the prophetic visions of St. John and to the second coming of Jesus Christ. Also known as the "Apocalypse," from the Greek for "unveiling," this book contains fifty-six separate prophecies, many relating to the imminent onset of a one-thousand-year reign of Christ, the millennium, followed by the final resurrection, the final judgment, and the end of the world. Prior to the institution of this reign, all kinds of misfortunes are predicted to befall the earth, culminating in the great battle at

Armageddon. These predictions are outlined in chronological order in Table 1.

The Bible has served as a direct source of prophetic prediction primarily for Jews and Protestants. The Catholic point of view as expressed in the Catholic Encyclopedia is: "Christian revelation has nothing to say about the end of the world as a purely physical phenomenon that can be forecast or described in scientific terms. To seek such information in the Bible is a waste of time." However, a large number of Catholics throughout history and to the present have reported apparitions and visions of divine personages and have transmitted predictive messages ranging from the detailed personal to national and global topics.

PROPHECIES OF CATASTROPHES

The prediction of dire happenings, disasters, and catastrophes has always been a special preoccupation of prophets. In fact, a close examination of the literature of any of the subjects discussed above will reveal that most of the recorded predictions deal with personal or national calamities, while only a few give grounds for optimism. Each period of history as well as every nation and culture have had its prophets of doom who have forecast disasters for the immediate and the distant future. In this section we will merely take a brief look at some of the catastrophe predictions that are operative in our time.

Natural Catastrophes. Astrologers, evangelists, and mystics issue a steady stream of earthquake alarms for most parts of the globe. Often these coincide with warnings by seismologists. As a result there have been numerous "correct" unscientific predictions of every major earthquake; but there appears to have been no systematic attempt to keep track of all incorrect or unfulfilled forecasts.

For example, in 1989 a group of Catholics in New York City reported that they saw the Virgin Mary who warned them of "our

TABLE 1. Apocalyptic Predictions from the Book of Revelation[a]

| Major references | | Events | Other references (inserted in Revelation) |
1st cycle Chs. 1–11	2nd cycle Chs. 12–22		
		A. John's Own Time	
	12–13	Christ's birth and ascension; Satan is cast to earth but persecutes the church during Roman empire	
1–3		John, imprisoned on Patmos, writes to the seven churches	10–1:2:22:6–21
		Fall of the Roman empire, A.D. 476	17–19:5
		B. General Matters Continuing Today	
	12:17	Satan continues to persecute the church	7:9–17
4–6:11		God in heaven; aggression, war, famine, death, and martyrs on earth (the part = "the great tribulation")	11:3–10
		C. God's Wrath, Immediately Followed by Christ's Second Coming	
6:12–7:8	15–16:9	The saints are "sealed" so as not to be hurt; then comes a great earthquake and four universal disasters	8:2–13
6:16, 8:11	14:14	Christ's glorious second coming	1:7; 19:11
	14:1–7	The first resurrection (the saved dead) and the rapture of the living church to be with Him	20:4–6
			11:11–12; 19:6–10
		D. God's Wrath Against the Unrepentant After Christ's Coming	
9 11:13–15	16:10–21 19:11–21	The last three woes: Jerusalem suffers, but the Antichrist is destroyed by Jesus at Armageddon	14:17–20
		E. Christ's Reign on Earth and the Final Events	
11:16–17	20:1–6	The Lord's millennial kingdom	
11:18	20:7–15	The final rebellion; the final resurrection (including the lost) and the final judgment	
11:19	21–22:5	The new heaven and the new earth	

[a]From J. B. Payne (6) by permission of HarperCollins Publishers.

planet being struck by a 'Ball of Redemption,' a fiery comet of Divine origin." The prophecy went on to state ". . . Another area that shall be shaken will be California. There is a great split in the earth that is widening." In June 1989, this same audience announced that Jesus communicated to them that "There will be earthquakes in many places. The present ones have been nothing compared to what will happen next. There will be a great earthquake in the Los Angeles area and also New York."

Global Wars. Along with all his other vaticinations, Nostradamus also foresaw World War III. According to Rene Noorbergen's[7] interpretation published in 1981: ". . . two hundred and sixty-five quatrains of Nostradamus [that] describe a worldwide holocaust; a war which has yet to be fought. The conflict, according to the seer, will involve not only the United States and Russia, China, the Middle East and Latin America but many other areas. In fact the entire world will be involved, with nuclear bombs, bacteriological weapons and gas killing off the population of country after country. . . . The time for this global clash has been set by Nostradamus as somewhere between the 1980s and 1995. . . . It will be the change in Russia from Communism to a more tolerable form of government that will be the catalyst. Nostradamus predicts that this change in Russia will open the way for friendship between the United States and Russia. . . . [this] will disturb the balance of power . . . the peoples of the Far East and the Middle East will retaliate." This is followed by a detailed interpretation of the 265 quatrains, which provides a blow-by-blow description of the global conflict. The part of this prediction about changes in Russia seems remarkably correct. We will have to wait to see about the dire consequences.

The Apocalypse. Since the earliest days of the Christian era, virtually every generation has had it share of prophets and charismatic leaders who proclaim that the end is near—the end prophesied in the Book of Daniel and the Book of Revelation of the Bible. The overall message of Revelation is very good news, at least for the Christians who will dwell in Christ's kingdom on earth and eventually proceed to Heaven.

However, the second coming of Christ is prophesied to be preceded by the worst of catastrophes: major earthquakes, famines, plagues and pestilence, and calamitous wars culminating in the great battle at Armageddon. These are so graphically described in the Bible that they have always fired the imagination of the devout and of nonbelievers alike. This is exemplified by Albrecht Dürer's famous etching, shown in Figure 2, which pictures the terrifying Four Horsemen of the Apocalypse that symbolize the chaos and misery to come. So pervasive and vivid are these biblical prophecies that the term "apocalyptic" has acquired a broader and more general meaning, suggesting any definitively catastrophic event.

However, a number of Christian sects, the millenarians, take the biblical prophecies literally and preach the imminent onset of the prophesied apocalyptic events. In his seminal book, *Pursuit of the Millennium*, Norman Cohn[8] surveys the millennial movements of the Dark and Middle Ages, while Michael Barkun[9] extends the analysis to the modern era. Some of the more prominent millenarian movements today include the Seventh Day Adventists, Jehovah's Witnesses, and many fundamentalist Protestant sects. Many of these have moved close to the American mainstream and publicize their predictions of imminent catastrophes in all of the mass media. And, of course, Nostradamus was not silent on the subject of the Apocalypse either. One of his quatrains states:

> The year 1999, seventh month,
> A great king of terror will descent from the skies,
> To resuscitate the great king of Angolmois,
> Around this time Mars will reign for the good cause.

Stewart Robb[5] offers the following interpretation: "In this remarkable quatrain, remarkable because Nostradamus not only mentions air-warfare but places it in the 20th century—we see that Armageddon is on in 1999. Mars here obviously signifies war. Its

FIGURE 2. The Four Horsemen of the Apocalypse by Albrecht Dürer (courtesy of the Fogg Art Museum, Harvard University, Cambridge, Massachusetts, gift of Paul V. Sachs).

beneficent aspect—'for the good cause'—means that the war is nearly over, with victory on the side of right, i.e., those fighting for the 'good cause.' [The original French reads 'bonheur' which can mean good cause, good time, or happiness.] Now, according to the Pyramid prophecy, the millennial dawn is due in 2001. So if Armageddon is nearly over in 1999, and over by 2001, its actual termination should be 2000 A.D."

THE COMMON THREAD

The methods for prophesying and predicting that are briefly described above have all been around for thousands of years. And similar ones have long existed in Asia and Africa and to some extent among Native Americans. To anthropologists they suggest the existence of a very basic component in the makeup of all human beings and societies—the need somehow to come to grips with the overwhelming forces that terrorize people, nations, and humanity at large, the quest for the supernatural.

The term "supernatural" is defined as "something caused by other than the known forces of nature." As such, the supernatural is a moving target. In primitive cultures there is little knowledge of natural phenomena, so the gods are credited with governing almost every aspect of nature and of human affairs: the motion of heavenly bodies, the tides of the seas, the change of the seasons, birth and death, and so forth. As many of these occurrences become better understood, they are gradually shifted from the supernatural to the natural. But in the populace there always remains a lingering awe for the unknown, the unexplained. And the future is, of course, one of the big unknowns.

All of the supernatural approaches to prediction are based on the premise that the future is largely, if not completely, determined. God or the gods have decided in advance the outcome of all major events or at least predisposed things so as to favor certain outcomes. By and large, the future is etched in stone. And the role of the chosen, the initiate, the prophet, or the seer is to read these etchings and to communicate them in the proper fashion. The recipients of these forecasts may prepare themselves and in some instances take advantage of options, but in general there is no escape from the long arm of destiny. Rarely do the questions "why?" or "how?" arise. These are usually considered irrelevant and pointless: que sera, sera; whatever will be, will be.

All who make use of the supernatural to forecast worldly events make one big demand of their audience—faith. Faith in the presumed source of the information and faith in the person transmitting the information. In biblical prophecy, for example, the Bible must be accepted as the word of God, an expression of the absolute truth, on faith. Elaborate systems of logic can be used to make inferences from what is written in the Bible, but the fundamental authority of the Bible has to be accepted without argument. Where modern prophets use the Bible to make predictions, their authority must also be accepted. The same applies to seers using magic, astrology, or other occult or "unscientific" methods. If the authority of the source and of the prophet is accepted on faith, everything else falls into place. If not, the prophecies lose every shred of credibility.

My own attitude is one of aloofness, of detachment. I am a scientist. I try to avoid being overly "unscientific" in running my personal life, and I certainly would not interject the supernatural in any of my professional work. But, as I later explain, I fully recognize that science and the scientific method also rest on fundamental axioms and assumptions that must be accepted on faith. I have great admiration for the past achievements of science, and I have confidence in the ability of scientists to devise solutions to many of our problems. Nonetheless, I accept that there are other systems of knowledge, other worldviews, that have some merit. I just keep my distance from them. But even so, I have been known to cast a furtive glance at the astrology column of the daily newspaper.

UNSCIENTIFIC VERSUS SCIENTIFIC PREDICTIONS

In the course of the past two hundred years, scientists have succeeded in carving out a respected and dominant position in the eyes of governments and of the public. Most decision makers turn to scientists for predictions of problems to come and for advice on how best to handle them. Certainly there are individuals in the government and industrial hierarchies who consult unscientific, even occult, prognosticators; but they generally do so surreptitiously. Science is in command; and yet, the thin line that separates the scientific from the unscientific is often far less distinct than most scientists would care to admit.

Scientists tend to scoff at unscientific predictions, because they are so laden with obscure symbolism, so vague, and so imprecise. In most instances, the true meaning of such a prediction becomes clear only after the fact, after it has been fulfilled. That is certainly a valid criticism. On the other hand, a scientific prediction that "there will be a major earthquake in southern California within the next fifty years with a 35% probability" is not exactly a paragon of precision either.

Unscientific predictions are also criticized because they are so subjective, relying as they do on intuition, insight, and a personal relationship between the prophet and the supernatural. In fact, intuition, inspiration, and luck often also play a vital role in scientific predictions. However, here there is a big difference. In unscientific predicting, each prophet is more or less on his or her own. There are very few useful tips that can be passed on from one prophet to the next. As a result, the quality of unscientific predictions has not changed materially over the past two thousands years. The prophets of old were no worse than the prophets of today. Some would say that they were better.

By contrast, contributions in science build on each other. A scientific advance starts out with an ingenious idea on the part of an inventive individual. But then a host of other scientists join in, many inject their own ideas, and there is continuous progress. As we will see in the next chapter, all predictions of future events, scientific or unscientific, entail the use of models. In this context, a "model" is any thought process, rational or irrational, by which future events can be predicted. The models used by scientists are based on causality, on a detailed understanding of the reason that events occur as they do. The more complete this understanding, the better is the prediction. So as scientists study natural phenomena, their understanding of what they observe improves, their models become more correct, and their predictions become more and more accurate—up to a point.

Science owes its commanding position in our modern culture in large part to the skill of scientists in formulating models and in guiding the evolution of the models over time as more and more knowledge becomes available. At the start, a scientific model is merely a hypothesis—a conjecture, a suggestion, a scenario. Then the model is tested to see how well it predicts events that can be observed, and if necessary, it is gradually improved on the basis of these observations. This process is continued until the model proves sufficiently reliable, sufficiently *valid*, to permit its confident use to predict future events. Without this kind of validation the predictions are little more than guesses, educated guesses, but guesses just the same.

As we will soon see, when scientific models are used to predict catastrophic events, phenomena of extraordinary magnitude that have never happened before, adequate validation is not possible. The predictions then reflect the scientists' personal insights, biases, and arbitrary decisions. They are framed in the jargon of science, but they are not *scientific* predictions. Ultimately, they may be no more meaningful than predictions made by some of the unscientific methods discussed here.

THREE

Scientific Models

THE RISE OF SCIENTIFIC PREDICTION

In the preceding chapter we saw that the prediction of important events in general and of catastrophes in particular is almost as old as human history. To that end, the prophets and seers of old employed religious insight, astrology, magic, and other occult means. Many of these practices persist to the present day. While the precise manner of generating such predictions varies widely, almost all share one underlying premise: they assume predestination—that the future is predetermined, that it is etched in stone. And many people around the world still feel that way and continue to seek the advice of clairvoyants, seers, and prophets. But over the past two or three hundred years an alternative worldview has emerged: science.

The basic idea of science is that we can come to understand the world by reasoning about it. By observing the world around us and by using our intellect we can explain why things happen the way they do. We view events, including catastrophes, as the direct result of a chain of circumstances or causes. And we can hope to understand that chain and perhaps to modify it by timely intervention. This is not an obvious concept. There are no stone tablets from Mount Sinai to assure us that we can get anywhere by starting from observations and using logical inferences. That is an assumption that cannot be proved, one that must be accepted on faith; the entire edifice that is modern science is based on that premise. Some would argue that science, therefore, is really a theology and that it is no more and no less firmly founded than most religions.

It took a long, long time for science and its practices to be permitted, let alone accepted. The basic premise of science was contrary to the teachings of the revered Aristotle and most of the respected philosophers. Science ran counter to the prevailing dogma that all knowledge is revealed, for example, in the Bible, and for that reason was bitterly opposed by the Church. Early scientists such as Copernicus and Galileo were regarded as iconoclasts and heretics. Others tried to compromise by attempting to fit the scientific approach into the prevailing religious context. It was not until the nineteenth century that scientists began to be taken seriously by rulers and decision makers, often only on an "if all else fails" basis. Their influence grew gradually during the last century as a direct result of impressive achievements in medicine, agriculture, physics, chemistry, and the like. Eventually, the scientific approach to dealing with personal and social issues assumed a dominant position in most countries—to the point that in our day it is an enormous embarrassment when a head of state is revealed to have consulted an astrologer. The scientific method thus prevails. We will now explore the worldview of modern scientists and the way that they predict the flow of events in the future.

THE SCIENTIFIC METHOD

A scientist begins by using his senses or his instruments to observe some phenomenon that interests him. From this he endeavors to form a scientific description or explanation of that phenomenon, one that will permit him to predict things that he has not observed or measured. There are many areas within science, and each has developed practices that are to some extent unique unto itself, but the general methodology for doing this is firmly established and is called the *scientific method*. This entails the following four steps.

Data Gathering. The scientist selects some aspect of the natural universe and collects much information about it. Preferably this is done with forethought in a well-organized, systematic manner so as to facilitate the next step.

Hypothesis—Model. The scientist ponders the data that he has collected and uses all the insight, intuition, and ingenuity that he can muster to make a tentative generalization—a hypothesis that serves to explain why the data take the form that they do. This hypothesis is often called the *model* and consists of a set of concepts that is meaningful and useful in understanding the data that have been gathered.

Prediction—Validation. The scientist now uses the model to predict something new. In this context the term "predict" does not necessarily mean the forecasting of some future event, though it might. In fact, the scientist may have deliberately set aside some of the observations that he has gathered just to find out how well his model is able to "predict" these data; or he may use someone else's observations made in the past; or he may actually wait and use data that he gathers after he has formulated his model. In any event he endeavors to "validate" his model by comparing the predictions made using the model with those actually observed. The more successful the predictions, the greater the confidence that the scientist has in the model and the greater is the validity of the model.

Refinement—Iteration. Usually a model as originally hypothesized is unsuccessful or only partially successful in predicting events. The scientist must then head back to the drawing board and try for a better model. Perhaps more and different observations need to be gathered. Perhaps new ideas are required. Perhaps some changes or fine tuning of the original model will do the job. In any event, the above steps are repeated until an acceptable model results, or until the scientist gives up.

Consider the classical, possibly apocryphal, story of Isaac Newton. He supposedly sat under an apple tree and observed a falling apple (data). In a flash he got an idea: masses attract each other. He thereupon hypothesized the law of gravity (model). When he expressed this law mathematically, it was used to "predict" the orbits of planets, and these predictions were confirmed when compared with the observations that he made with his telescope (validation). Presto! Of course, it is rarely that straightforward, and many false or partially false hypotheses may be made before a satisfactory model is found. And only very rarely is a model so significant and valid that it is given the honorific title of "law." A *law* is a model that is accepted as valid by the members of a scientific discipline. It is treated as a fact; but even laws are occasionally, though rarely, disproved or modified by new scientific evidence.

WHAT IS A SYSTEM?

No scientist would dream of trying to study or describe the world as a whole. Instead he deals only with certain well-defined aspects of a very limited portion of the real world. Usually his training and professional experience have been focused on that tiny corner of reality. If his field is aerodynamics, for example, his concern may be how a specific kind of airplane, or perhaps a bird, manages to fly. He is not interested in the appearance of the plane, nor its cost or its destination, just in how fast it moves and in what keeps it from falling down. The term *system* is used to designate the specific aspects of the plane in which he is interested.

Few words in our modern technical lexicon have been more vulgarized, hackneyed, and deprived of meaning than the word system. We speak of philosophic systems, systems for winning at the horse races, systems of classification, etc. In many discourses, the term is used as a synonym for the word "something." We need to be more precise than that. I will use the term "system" to designate that part of the real world that is under examination and the behavior of which we want to predict. In Figure 3, the system is represented as a box, labeled S, with a number of inputs, labeled I, and a number of outputs, labeled O. In system terminology, the inputs and outputs are variables, meaning that they are quantities that can be measured or counted or otherwise described by measurements. The inputs are a set of variables that describe the injection of matter, or energy, or some other measurable quantity into the system; and the outputs are another set of variables that are of particular interest and importance to the scientist; he may wish to predict some of them.

In the kinds of systems that we will discuss, these outputs generally vary with time, assuming different magnitudes at different instants of time, and are termed *transients*. For example, the transient temperature of the human body varies throughout the day. We also say that the temperature is a "function" of time. The inputs or input variables are also capable of being measured and are frequently transients. They play an important role in determining the output variables. In mathematical terms the system "maps" or transforms the inputs into outputs. Inputs and outputs are also frequently called the *excitation* and the *response*, respectively.

In the case of aircraft dynamics, for example, the input variables might include the thrust exerted by the jet engine as well

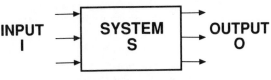

FIGURE 3. System schema.

as the manipulations of the aircraft controls by the pilot. The output transients would include the resulting speed and the direction of motion of the plane.

In biology, to cite another example, the system S might be the endocrine system of a laboratory animal, the input I might be a new drug fed to the animal at intervals, and the outputs O might be resulting changes in blood pressure, pulse rate, and chemical measures of the metabolism. In economics, the system might be the market for some commodity, the inputs would include measures of the supply and the demand, and the outputs might include the price of the commodity.

So, in general, in dealing with systems we have: the system, the inputs, and the outputs. Typically, the scientist will know or be given two of these and try to find the third. There is, in fact, a technical specialty that deals with the problems arising in the study of systems: systems science. The problems confronting a systems scientist fall into three broad categories: analysis, synthesis, and control, depending on whether the system, the inputs, or the outputs are to be determined. This is shown in Table 2.

In the system *analysis* problem, descriptions of the inputs and the system are provided, and the scientist is asked to find or to "predict" the outputs. For example, in the case of the airplane, he might be given a description of those parts of the plane that relate to its dynamics and asked to determine how the plane will respond when the pilot pulls back on the stick. Or in the case of the analysis of an electrical system, the scientist might be given the circuit diagram of a toaster and be asked to predict the transient

TABLE 2.   Classes of System Problems

Type of problem	Given	To find
Analysis	Input, system	Output
Synthesis	Input, output	System
Control	System, output	Input

current and the transient temperature that would result from plugging in the toaster and turning it on.

In the system *synthesis* problem, the scientist is given a set of inputs and a set of outputs, and he is asked to find or design a system that responds with the specified outputs when excited by the specified inputs. For example, he might be asked to design a plane having a desired maneuverability. Or in the case of the toaster, he might be asked to design the electrical circuit of the toaster so that the temperature reaches a set value within a specific time period.

The third class of system problems, the *control* problems, involve the specification of the system and the desired outputs, while the inputs are to be found. For example, the scientist might be given a description of the airplane and asked to find out how the aircraft controls should be manipulated, perhaps by an auto-pilot, so that the plane will maintain its heading (output) when buffeted by wind gusts.

In comparing the three classes of system problems, we find that analysis problems are generally easier to solve than synthesis or control problems, because only analysis problems have unique solutions. Taking the toaster as a simple example, we can use the simple rules of circuit theory to find the current that will flow into the toaster when it is plugged in. By Ohm's law, that current is equal to the voltage source (say, 110 volts) divided by the resistance of the toaster, a value in ohms provided as part of the description of the system. There is only one correct answer, a unique solution. This is generally true of analysis problems. Exceptions exist (as we will see when we discuss the chaos phenomenon in the next chapter); but it holds well enough for the purposes of the present discussion.

By contrast, synthesis and control problems never have unique solutions. In fact, they always have an infinite number of possible solutions. For example if asked what electric circuit will draw a current of one ampere when connected to a one-volt battery, we might suggest a system consisting of a single one-ohm resistor. But we could also suggest two one-ohm resistors connected in parallel, or two half-ohm resistors connected in series. There are, in fact, an infinite number of resistor configurations that have the same input/output behavior and that are, therefore, correct solutions to this synthesis problem. To get a practical solution we must have a way of deciding which of the many possible solutions is the one that we really want. Perhaps the one that is the cheapest, or the lightest, etc. In systems science terms, we need to impose separately specified constraints, and that introduces many additional difficulties.

CAUSALITY

Fundamental to the systems approach is the concept of causality. The system S in Figure 3 is said to be causal if the outputs are caused by the inputs. This may at first seem to be a very simple notion, but it actually embodies a wealth of subtleties and nuances that have kept philosophers busy for hundreds of years. Let's see what our intuition suggests to us.

Looking at Figure 3 and thinking of the examples from aerodynamics, biology, or economics that were briefly presented above, we can say that the outputs should succeed or follow the inputs—at least they cannot precede them. But must the outputs *always* follow the inputs? Not really. We may wish to allow for occasional lapses or exceptions and require only that the outputs follow the inputs most of the time, that there be a statistical correlation between inputs and outputs. For example, that if we predicted that a certain output would follow a certain input, we would be correct say 99% of the time. But would that prove causality? Consider the following example.

We are in the habit of taking a train to town every weekday at 9:00 A.M. We observe that the train pulls out of the station each morning precisely when the big station clock shows nine o'clock. It happens again and again. We can count on it but is there a causal relationship between the clock and the train? Had we the opportunity and interest, we might try an experiment and sneak into the station some night and advance the clock by two minutes. If the train left two minutes early the next morning, we would have some interesting evidence. But the very next morning the conductor might decide that the clock was no longer trustworthy and use his watch to determine the departure time and there would go our proof of causality. The point is that it is extremely difficult, if not impossible, to tell simply by observing a system whether there really is a causal relationship between two events that we regard as input and output. Do they, in fact, play the roles of excitation and response, or are both events really responses caused by an input we haven't thought of.

Actually, the only way that we can be confident that we have established the existence of a causal relationship is to provide a plausible explanation of why and how the inputs cause the outputs. In the above example, that might entail interviewing the conductor or a railroad administrator to learn for sure who decides when the train is to depart and how he makes his decision. This is true for all systems. The only way that we can be confident of a causal relationship between inputs and outputs is to propose a plausible causal chain linking the presumed inputs to the observed outputs. In the train example, such a chain might include the following links:

1. The conductor realizes that it's nearly nine o'clock.
2. The conductor looks at the station clock.
3. When the conductor sees that the big hand has reached twelve, he blows his whistle.
4. When the engineer hears the whistle, he starts the train.

We can convince ourselves even more by watching the conductor and the engineer in order to check out the above four steps. We might be satisfied with such an explanation because it makes sense to us; it is plausible. It conforms with the way we believe things to work. The term *model* is used to designate the causal chain that characterizes or explains the behavior of the system.

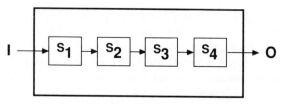

FIGURE 4. System with four subsystems.

The model for the train example is illustrated in Figure 4, which is similar to Figure 3 but shows the system S to be comprised of four subsystems: S_1, S_2, S_3, and S_4, which represent the four steps. The system S had "disaggregated" into a more detailed model consisting of four elements—the causal chain. It may be important to us that the model we constructed is the "correct" one and not just a reasonable one, because we may want to use the model to predict other input/output relations. For example, we may want to predict whether the ten o'clock and the eleven o'clock trains will depart when the big hand of the clock reaches twelve.

And so it goes with systems in general. In order to satisfy ourselves that the system meets the causality condition, in order that we can use its model to predict outputs, we need to look inside the system. Inside we find a lot of other systems connected together as a simple chain as in Figure 4, or as a simple network as shown in Figure 5, or in much, much more complex ways. Actually, each subsystem, each system within a system, may itself be represented by a network of subsystems, and each of these subsystems could be further disaggregated, and so on.

In the train example we might decompose subsystem S_3 to describe the sequence of actions encompassed by the step: "3. The conductor blows his whistle." These might be: he takes the whistle out of his pocket; he puts it into his mouth; he blows into it; etc. In general, we believe that we have demonstrated causality, and therefore the utility of the model to predict outputs, if we can find a continuous path from input to output via the subsystems and if

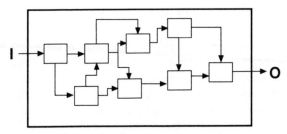

FIGURE 5. System with a network of subsystems.

we consider the causal chain that we describe in this way to be plausible.

As a much more complicated example of the systems approach, consider medicine as it is practiced conventionally. A patient comes to a physician with some specific complaint, perhaps a pain in his side. The doctor has been trained to view the human body as a network of subsystems, such as the nervous system, the endocrine system, the respiratory system, the various organs, etc. Each of these is viewed as being comprised of small subsystems. Using his diagnostic techniques, the doctor identifies the subsystem (or sub-subsystem) that is causing the trouble. He then attempts to eliminate the problem by treating the affected subsystem chemically or surgically. This is in sharp contrast to the holistic healer who might prescribe prayer, positive thoughts, exercise, diets, or more exotic methods of treating the body as a whole.

We know, of course, that on occasion the conventional physician will fail to cure an ailment and the holistic healer will succeed. However, it is the position of most mainstream scientists that such a setback is temporary and that a satisfactory cure for the subsystem can eventually be found. This will probably involve a revision or a refinement of the model and may take a lot of time and money; but there is unbounded confidence in the systems approach and the scientific method. The important thing is to find the right model.

PLAUSIBILITY

We now return to the problem of causality. We just stated that a necessary step in establishing that a system is causal is to provide a plausible chain of circumstances or events—a chain linking the inputs to the outputs. In terms of the three classes of system problems, we are faced with a synthesis problem. We are given the inputs, I, and the outputs, O, and we are asked to find the system, S. Such a problem does not have a unique solution. There are many possible models—many causal chains that could be suggested to account for the behavior of the system. But there is only one "correct" one, only one that can be used with confidence for prediction.

We can never be absolutely certain that we have found the "right" solution, the "correct" model. But we can reduce our uncertainty by following time-tested procedures in designing our model, procedures that have worked well in similar circumstances in the past. Of great importance is that we must insist that the model be plausible, that it conform to our general knowledge and expectation of how the model should look. But we must now ask: Plausible to whom?

The model, of course, should appear to be reasonable to the person designing it and also to his close collaborators. But that is not enough if the model is to be used to predict future events and if it's hoped that others, perhaps the public at large, will accept the predictions. For that, the model must have *credibility*. And to be considered credible, the model needs some sort of seal of approval. Anyone can design a model and make predictions by scientific, pseudoscientific, or unscientific methods. How is one to know if the person doing the predicting is intelligent, reliable, and honest? In our modern science-oriented society, a model is considered credible only if it has the support and approval of a respected body of specialists and experts. Explicitly or implicitly the government and the public vests such a body with the authority to determine whether a model, and the predictions that are made using it, should be accepted and used. Now just who are the people that comprise that body of experts?

When the scientific method is applied to a new area, one that has not been carefully studied previously, this body is usually a small, tight group of scientists and researchers. If their modeling efforts meet with some success and if the model has significant and broad applications, more and more scientists will join in. Eventually, a new scientific or professional society may be formed, followed by new scientific journals, academic departments, and college curricula until, perhaps, it is recognized that a new scientific discipline has emerged. Somewhere along the way, some of the people providing the money to pay for the research to improve the model and some of the people who are making serious use of the model join in, and a new Community may be formed.

In 1962, in his very influential book, *The Structure of Scientific Revolutions*, Thomas S. Kuhn[1] introduced the use of the term "paradigm" to characterize the essence of scientific disciplines. These paradigms include the basic models that have been successful and that have been accepted, as well as the methodology and practices used to derive or design these models. A very important function of each discipline is the nurturing and safeguarding of the paradigms to assure their continued integrity. Members of a discipline jealously guard its paradigms from unprincipled or uneducated interlopers, an activity that sometimes gets in the way of progress. It may, in fact, require a "scientific revolution" to gain acceptance for a radically new and different model. On the whole, however, the scientific disciplines serve as the "certification" agency for the models and predictions that fall under their aegis.

The notion of science-based advocacy Communities was introduced earlier. These Communities often assist in establishing the credibility of models. There is an important difference, however, between a discipline and a Community. Scientific disciplines are established to expand our knowledge and understanding of interesting phenomena and are expected to thrive and flourish indefinitely or at least until they are "overthrown" by a scientific revolution. By contrast, Communities have a much more narrow goal or *raison d'être*. They are formed to answer a specific need, often one resulting from the emergence of a new problem or crisis, and they fully expect to dissolve and to disappear when that problem is solved in one way or another. Communities are therefore less permanent and less authoritative than scientific disciplines. However, they often play an important auxiliary role in the achievement of credible models.

In any event, it is the existence of an active and viable discipline that is the key to the establishment of the credibility of models. A model that has been designed in accordance with the paradigms of the discipline is warranted to be plausible and to merit the adjective "scientific." By implication, a model that does not conform to the paradigms of some established discipline is given labels such as "unscientific," "pseudoscientific," or worse. And all members of the scientific discipline are expected to conform in their choice of models and paradigms.

An example: The significance of all this came home to me early in my career. In the early 1950s, I was working for a geophysical company engaged in the modeling of geologic formations in order to locate underground oil pools. During the latter part of the Korean War, the company received a contract from the Department of Defense to explore "exotic" methods for the detection of land mines. Land mines are explosives housed in metal containers, concealed underground, and set to explode when a person or a vehicle compresses the earth overhead. Conventional land mine detectors are actually metal locators—instruments mounted on the end of a long pole that sense the proximity of the metal container and alert the infantry man that there is danger ahead.

However, when the Chinese entered the Korean War, they brought with them new weapons: land mines in wooden rather than metal boxes. These were low-quality affairs and were probably introduced because of a metal shortage in China. But they proved to be a major headache for the U.S. Army. Conventional land mine detectors did not work because there was not enough metal to detect. Hence, the search for "exotic" methods. I became the project leader and eventually learned what the Pentagon had in mind.

Someone in Washington had learned of dowsing, an age-old technique for locating lost objects, underground streams of water, oil reservoirs, and other hidden objects. A *dowser*, sometimes called a water witch, is a gifted person who walks over land areas with a Y-shaped contraption in his hands looking for things beneath the earth. Some dowsers use forks of tree branches, others use bent coat hangers, and still others keep the nature of their "dowsing rod" a secret. When the dowser passes over the object of his search, the dowsing rod that he grasps with both hands starts to shake uncontrollably—the indication of a find. Dowsing is a respected profession in many parts of the world. Farmers often hire dowsers to determine the best locations for wells; oil companies have used them on many occasions; archeologists use them to find ruins of old civilizations. I understand that dowsers accompanied the British Eighth Army during its World War II desert campaigns and were charged with finding water for the troops. More recently, a major computer manufacturer employed a dowser to locate several wells near one of its California factories.

Anyway, the Army reasoned: "If water and oil, why not land mines?" So we set up a dummy mine field at any Army base in Virginia and advertised for volunteers. In directing the project, I met some fascinating people and had some unusual experiences. But it gradually dawned on me that if I persisted in this work, I would soon lose my credentials as a serious and ambitious scientist. Research into dowsing was definitely off limits for anyone who wanted to belong to the scientific establishment. The reason? Nobody has succeeded in formulating a plausible causal model linking objects underground (inputs) to the shaking of the dowser's hands (output). It was fairly easy to perform experiments to eliminate from consideration all "natural" phenomena that came to mind: electromagnetic radiation, smell, acoustic signals, telltale visual clues, and so on. The only thing that remained was "extrasensory perception" (ESP) and that was considered synonymous with hocus pocus. Absent a plausible causal model backed by an established scientific discipline, watching people with forked sticks in their hands was considered pseudoscientific at best. So, even though I was fascinated by the problem and well paid, I quit and moved on to other more solidly founded projects. The point is that modern science places a high premium on conformity with the paradigms of a discipline. Inventiveness and originality are encouraged, but only within the walls that each discipline has carefully built up in order to protect itself.

Incidentally, in the intervening years certain sectors of the scientific establishment have loosened up a little, and some studies of ESP and other so-called parapsychological phenomena and the like are finding a measure of support. But, on the whole, the same stricture is still operative: no causal model, no science.

MODELS AS ABSTRACTIONS OF REALITY

As the term is used by scientists, a model is an abstraction of some of the attributes of a specific object. It's an abstraction or condensation because even the simplest of objects has a myriad of attributes, and it is up to the modeler to decide which small subset of these characteristics to include in his model. Consider an ordinary lead pencil. We could use a model to describe its mechanical properties—hardness, flexibility, density, etc. Or we could model its optical properties—shape, color, reflectivity, etc. Or we could model its utility—the purpose for which it is designed and how it should be used. Or we could model it from the economic point of view—manufacturing cost, wholesale and retail price, etc. Each of these models would constitute an alternative, different representation of the pencil. Each of these models is an abstraction because it includes only a very small number of the characteristics or attributes of the pencil. These models do not conflict with each other; there is no competition between them. They simply provide different answers to different questions.

There is some justification in considering all of our intellectual activities to entail modeling, since we could never think about any object or problem in its entirety. We always focus on one or a few very specific aspects. In sensing or observing real-world objects, we must filter out almost all the information, all the sense data that come our way, and absorb only those that are of direct interest at the moment.

Experimentation always involves modeling. How else could we decide what quantities to measure, where to place our instruments, how frequently to record outputs, etc. We must first have a model in mind and then use the experiments to refine it or to answer some specific questions. Likewise, decision making always requires models that allow us a simplified view of alternatives, as does every kind of prediction, whether by extrapolating from the past to the future or whether we have some sort of formula. Of course, most of the models that we use for everyday thinking are not scientific models; far from it.

Many disciplines have adopted mathematics as the preferred language for expressing their models. One reason for this is that by using the strict rules of mathematics, most, if not all, of the possibilities of ambiguities that exist in natural languages, such as English, can be eliminated. Another reason is that as the systems being modeled become more and more complex, the determination of system outputs becomes more and more difficult and time consuming. Where models are formulated mathematically, modern high-speed computers or supercomputers can often be used to find the desired solutions.

DEDUCTIVE AND INDUCTIVE MODELS

Valid models of systems are the key to the successful prediction of the response (outputs) of systems to specified excitations (inputs). It is now time to look more closely at the way that models are actually designed. There are numerous techniques, but all can be regarded as employing combinations in varying proportions of what philosophers call deduction and induction.

Deduction means starting with something general and deriving something specific. When we model deductively we begin with a scientific law, a broadly worded statement or generally applicable equation in which we have great confidence, and employ logical reasoning to arrive at the model. We recognize, of course, that fundamentally all of our knowledge is empirically derived from observation and sense data. No stone tablets, no Hermetic secrets, no sweeping intuitive insights are available to the scientist. From the scientific point of view, we are all born devoid of knowledge and only acquire knowledge through our senses. Everything that we know about the universe is based directly or indirectly on observations. So when scientists use the term *law* they really refer to a principle, based on observations to be sure, but one that has stood the test of time so long and so well that it can be accepted without question or further argument. It is something on which we can count. Where such laws are available, they provide fine foundations and excellent starting points for model building. We start with the law and deduce the model.

The basic alternative to deduction is induction. *Induction* means starting with specific information and inferring something more general. In inductive modeling we gather and accumulate much data—specific values of inputs and outputs. When we think we have enough data, we sit back and try to arrive at a generalization—a statement or perhaps an equation that applies to all or most of the data. Usually this entails discerning some sort of a pattern or a number of patterns in the data. There are many ways of recognizing these patterns, some very logical, others quite intuitive. Let's look at some examples of deductive and inductive models.

Deductive Models: A Very Simple Example. Consider the following problem from high school physics. A billiard ball is released from the top of a high tower (why not the leaning tower of Pisa?).

How far will the ball have dropped after three seconds? The system is illustrated on the left side of Figure 6, and the system schema is shown on the right side of Figure 6. In this case, the falling ball is the system S, the time of release of the ball is the input I, and the distance traveled is the output O. Galileo demonstrated that it does not matter how heavy the ball is. For the mathematical model we turn to the law of gravity, formulated by Isaac Newton around the year 1666. That law states that two bodies attract each other with a force proportional to each of their masses (or weights). Newton also proclaimed that the force on a body is equal to the product of its mass and acceleration. Putting these statements together results in the model, the equation

$$d = g \times t^2/2 \tag{1}$$

where d is the distance traveled in t seconds and g is the acceleration due to gravity, which is known to be about 32.2 feet per second per second. For any numerical value of t we can use the above equation, which is the mathematical model of the system, to find the corresponding distance traveled. For example, if t is 3 seconds, d turns out to be around 144 feet.

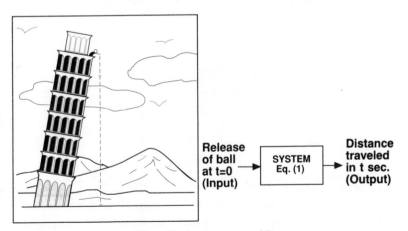

FIGURE 6. The leaning tower of Pisa.

Note that we did not have to go out and measure anything. We took Galileo and Newton at their words, and it is unlikely that we will ever be disappointed in doing so.

Deductive Modeling: A More Complicated Example. Consider the mechanical system shown in Figure 7, which consists of a metal block or mass M connected to a fixed support by means of a spring K. Think of a cheap bathroom scale turned upside down and glued to the ceiling. We will assume that we are told what the mass is and how stiff the spring is. The labels M and K can be used to represent, respectively, the size of the mass (its weight divided by the gravitational constant) and the stiffness of the spring (the force needed to stretch it a unit distance). Incidentally, M and K are called *parameters*. They are the attributes of the components of the system that help determine the relationship between system inputs and outputs.

We know that if we grasp the mass, pull it down, say a distance D, and then let go, it will bob up and down for quite a while. Suppose that we want to know how fast it will bob, how many times per minute it will move up and down. A more sophisticated way of asking the same question is to seek the

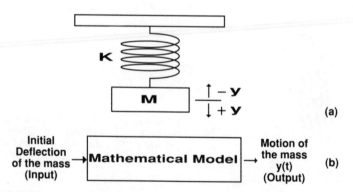

FIGURE 7. (a) Schematic diagram of a simple mass/spring system. (b) System schema of the mass/spring system.

frequency of the vibrations or oscillations of the mass. In terms of the systems terminology introduced in Figure 3, the system is as shown in Figure 7(b), the displacement D is the input, and the resulting motion of the mass is the output. If we use the variable $y(t)$ to measure how far the mass is from its equilibrium position (where it was before we displaced it) at any moment, the desired output we are looking for takes the form of a transient, y, as a function of time. Here is how we proceed to find the output.

We look in our physics textbook in the section that deals with mechanics for an applicable law. After a short search we encounter the ubiquitous Isaac Newton, who formulated what have come to be known as Newton's Laws. From these we can easily derive an equation that allows us to predict the displacement of the mass at all points in time after we release it. (This derivation is presented in the Appendix.) It turns out that the displacement y is given by the equation

$$y = D \cos(2\pi f t) \qquad (2)$$

which is a mathematical model of the mass/spring system. This expression describes simple harmonic motion, the way the mass bobs up and down. The f in Eq. (2) is the frequency, the number of times per second that the mass reaches the highest point in its trajectory. The frequency is proportional to the square root of the spring constant, K, divided by the mass, M. We can insert whatever values we like for the initial displacement, D, and for M and K to obtain the frequency of the vibration and the location of the mass at any time, t.

If the mass/spring system is accessible to us, we can attach a pencil to the mass and put it in contact with a sheet of paper that is slowly and uniformly moved along as shown in Figure 8. The pencil will then draw the curve that is expressed by Eq. (2). So we could actually find the frequency that we are looking for by setting up an experiment with an actual mass and an actual spring. But that is usually inconvenient and impractical. It is much easier and cheaper to work with the mathematical model and to solve it

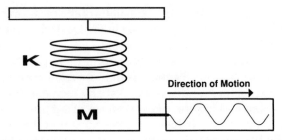

FIGURE 8. Experiment to determine and to draw the solution of Eq. (2).

without leaving one's desk. The model has another advantage. We can substitute any numbers we want for M and K and quickly obtain the frequency of the vibration. In this way we can quickly determine how the system of Figure 7 would respond if we used different masses and springs. If the problem were not so elementary, we might use a computer to find the solution of the equation or equations comprising the model.

Even if you have not followed the details of the above discussion, please take careful note of these points: We began with the basic laws of Isaac Newton, which apply to all mechanical systems. We then applied the laws to the specific system of Figure 7. In other words, we specified the structure of the mechanical system, and derived the mathematical model, Eq. (2). Finally, we inserted numerical values for the parameters, the coefficients of the equation, in this case for M, K, and perhaps D. The solution, Eq. (2), was obtained by solving a system analysis problem, by logical inferences and without needing to make measurements or to observe an actual system.

Such problems generally have unique solutions. There is only one correct answer. (We again defer discussion of the chaos phenomenon.) So we can be very confident that the output that we derived will be correct for any values of M, K, and D that we might select, provided only that the basic assumptions that we have made are correct. There are some caveats and pitfalls that we will discuss in Chapter 4, but it is generally the case that mathematical models derived deductively are relatively dependable because they are based on dependable laws and a solid causal chain linking the inputs to the outputs.

Inductive Models: A Simple Example. Suppose that in the preceding example we had been ignorant of the way a spring works. We would then be compelled to explore the behavior of the spring experimentally or empirically. We might set up a laboratory experiment, like the one shown in Figure 9(a), that allows us to stretch the spring and to measure the force required to do the stretching. Using this setup, we would stretch the spring a small measured amount (say, one inch) and record the numerical value of the force shown on the meter. Then we would stretch the spring by a slightly larger amount (say, two inches) and again record the force, perhaps on a graph as shown in Figure 9(b). We would keep doing this until we felt that we had enough data.

Looking at the graphical plot of the results of these measurements might suggest to us that the points more or less fall on a straight line. So we would draw the best straight line through the points. That would mean that there is a relationship of direct proportionality between force and displacement. And the slope of the line establishes the magnitude of that constant of proportionality. This slope is the parameter, K, that we could use in the mathematical model.

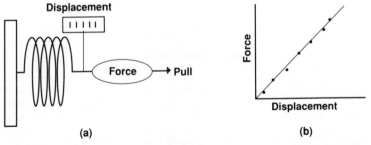

FIGURE 9. Experiment to measure the spring constant as a function of displacement.

By our experiment we have established the parameter K inductively. In effect we have solved a system synthesis problem, by starting with inputs and outputs to find the system. We can be confident of that solution only over the range of displacements that we have explored in our experiments, and it is only as good as our measurements.

Incidentally, it may happen sometimes that the data plotted in Figure 9(b) do not fall on a straight line. Instead they may suggest a curve. That would imply that K is not a constant but rather a function of displacement. The model would then be called a *non-linear* model, be more difficult to solve, and probably call for a computer solution. We will come to that later.

Inductive Models: A More Farfetched Example. Consider now an example taken from American history. Look at the following series of numbers:

$$. . . 1860, 1880, 1900, 1920, 1940, 1960, ?$$

It does not take much imagination to recognize that they represent years. Without further information, we would be tempted to say that the next number in the series should be 1980. But look at Table 3 which shows what these numbers represent. Each of the above years was a presidential election year in U.S. politics. And the president who was elected or reelected in each of these years died in office.

TABLE 3.   Fate of Certain United States Presidents

Year	President Elected That Year	Fate
1860	A. Lincoln	Died in office
1880	J. Garfield	Died in office
1900	W. McKinley	Died in office
1920	W. Harding	Died in office
1940	F. Roosevelt	Died in office
1960	J. Kennedy	Died in office

Actually, William Harrison, who was elected in 1840, also died while in office; but so did Zachary Taylor, who was elected in 1848. Anyway, as a number of people have pointed out, *every* president who was elected in a year that is a multiple of 20, from 1860 to 1960, died in office, and those were *the only* presidents to die in office during that period. Do you think that you see a pattern? If so, you might wonder why there was so much competition in the presidential election of 1980.

Seriously though, many people consider data such as that shown in Table 3 to be very significant and disturbing. They see a historical process or event repeated again and again and again. In essence, they perceive the model shown in Figure 10. What this says is that "Presidents elected in years that are multiples of 20 die in office." A more formal form of the same model would be:

Election in a year that is a multiple of 20 is a necessary and sufficient condition for the death of a president while in office.

Of course, Ronald Reagan broke the jinx. He was elected in 1980, and at the end of his second term in 1989 he walked out of the White House, smiling. He did have one or two close shaves though. So the formal model is not absolutely reliable, since there is one recent exception. Still, the model "predictions" look strikingly consistent.

What are we to make of this? If you were the director of the Secret Service charged with protecting presidents, would you have used this model as the basis for requesting a larger budget and staff to protect President Reagan? And would you make a

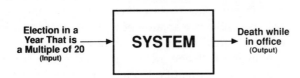

FIGURE 10. Schematic diagram of a model of presidential mortality.

similar request in the year 2000? Would you relax your vigilance until the year 2000?

If you are a hard-core scientist or if you have swallowed the arguments about causality presented earlier in this chapter, you would probably shout: "Get out of here! No way!" Or you might say: "Show me a terrorist organization that has it in for presidents elected in years that are multiples of 20, and I might take another look. Where is the causality? No causality, no credible predictions!" And you would have a good point, a point that could be elaborated by the following argument.

Make a list of all of the presidents, from Washington to the present, and next to each name place a long list of attributes—facts relating to that president: date of his birth, first and middle initials, name of wife or wives, number of children, first letter of name of birthplace, number of years in school, year first elected to public office, and on and on and on.

Table 4 is an example of such a listing; there is no special significance to any of the attributes that have been included. Is it not clear that if we make the list of attributes long enough, say 10,000 items, we could discover "astonishing" things? Take any subset of the presidents, any group of names from the list, and we could surely find some attribute that is the same for every member of that subset and that is different for all of the other presidents. Try it yourself. It's fun. For example, did you know that in the past 120 years only one president was reelected in a year ending with a "2"? And he resigned before completing his term. All we need is enough attributes. "Election in a year that is a multiple of 20" is just one such an attribute.

Table 3 is startling only because no other attributes are included; and that is the only reason that we might think that we discern a meaningful pattern, a significant model. Only if we have a convincing, plausible explanation of why an attribute might be significant are we justified in using it as part of a model. The model shown in Figure 10 does not include a plausible causal chain and is therefore not a scientific model. So the logical argument would run.

But wait a minute! Is that really fair? Most of the decisions that we are called upon to make day in, day out are based upon inductive models that have very little causal basis and are far less consistent than the data of the presidential mortality example of Table 3. We make decisions to questions such as "What shall I wear today?" "Should I carry an umbrella?" "Should I read this book?" "Should I even bother to get out of bed?" We are constantly compelled to make decisions like these, and we must make them intuitively with much less hard data than the presidential mortality example. Our intuitions are usually based on patterns that we may feel we have observed in similar situations in the past, but rarely do we have the time to look at things systematically.

Incidentally, suppose the string of numbers from 1860 to 1960 did not represent years at all, but rather house numbers on Elm Street. Suppose now that these houses, and *only* these houses on the street, recently had fires. Probably you would have a lot of trouble buying fire insurance for your home at 1980 Elm Street. The agent might be very cagey about letting you know the reason, but chances are that your application for insurance would be turned down. Only in very rare situations does an Isaac Newton come along and formulate a law on which we can base a deductive model.

Many people firmly believe in models such as the presidential mortality model. They make tables like Table 4 and massage the information in many ways. If they have computers they may use the statistical method called *regression analysis* to find correlations between suspected inputs and outputs. They do this to predict the stock market, to handicap horse races, to find the ideal mate, etc. Not infrequently, some of the predictions work out and some people are able to make a very good living selling their predictions. But they are not *scientific* predictions based on *scientific* models, and therefore few scientists have much to do with them. And neither do I. Still, had this book been scheduled for publication prior to 1989, I would have felt uneasy in using the presidential mortality example.

Hybrid Models. In most practical modeling situations, we encounter problems that lie somewhere between the purely deduc-

TABLE 4.   A Few of the Many Attributes of United States Presidents

	A	B	C	D	E	F
No.	Name	Year(s) elected	Terms or end	Party affil.	Birth state	Prof-ess'n
1	George Washington	1788, '92	2	Fed.	VA	mil
2	John Adams	1796	1R	Fed.	MA	law
3	Thomas Jefferson	1800, '04	2	R/D	VA	law
4	James Madison	1808, '12	2	R/D	VA	pol
5	James Monroe	1816, '20	2	R/D	VA	law
6	John Quincy Adams	1824	10	R/D	MA	law
7	Andrew Jackson	1828, '32	2	Dem.	SC	mil
8	Martin Van Buren	1836	10	Dem.	NY	law
9	William H. Harrison	1840	D10	Whig	VA	mil
10	John Tyler	1841	p, R	Whig	VA	law
11	James K. Polk	1844	1R	Dem.	NC	law
12	Zachary Taylor	1848	D10	Whig	VA	mil
13	Millard Fillmore	1850	p, R	Whig	NY	law
14	Franklin Pierce	1852	1R	Dem.	NH	law
15	James Buchanan	1856	1R	Dem.	PA	law
16	Abraham Lincoln	1860, '64	D10	Rep.	KY	law
17	Andrew Johnson	1965	1R	Dem.	NC	tailor
18	U.S. Grant	1868, '72	2	Rep.	OH	mil
19	R.B. Hayes	1876	1R	Rep.	OH	law
20	James A. Garfield	1880	D10	Rep.	OH	pol
21	Chester A. Arthur	1880	p, R	Rep.	VT	law
22	Grover Cleveland	1884	1D	Dem.	NJ	law

TABLE 4. (*Continued*)

G	H	I	J	K	L	M	N	
Age f'st elect.	Age el. presid.	Age at death	Date of birth	Zod. sign	Eldest son	No. of chil.	Death state	No.
26	57	67	02-22-'32	Aqu	yes	4	VA	1
33	61	90	10-19-'35	Lib	yes	5	MA	2
26	57	83	04-13-'43	Ari	yes	6	VA	3
25	57	85	05-16-'51	Pis	yes	—	VA	4
24	58	73	04-28-'58	Tau	yes	3	NY	5
35	57	80	07-11-'67	Can	yes	4	DC	6
29	61	78	03-15-'67	Pis	no	1	TN	7
30	54	79	12-05-'82	Sag	no	4	NY	8
26	68	68	02-09-'73	Aqu	no	10	DC	9
21	51	71	03-29-'90	Ari	no	15	VA	10
28	49	53	11-02-'95	Sco	yes	—	TN	11
64	64	65	11-24-'84	Sag	no	6	DC	12
28	50	74	01-07-'00	Cap	yes	2	NY	13
25	48	64	11-23-'04	Sag	no	3	NH	14
23	65	77	04-23-'91	Tau	yes	—	PA	15
25	52	56	02-12-'09	Aqu	yes	4	DC	16
20	56	66	12-29-'08	Cap	no	5	TN	17
46	46	63	04-27-22	Tau	yes	4	NY	18
42	54	70	10-04-'22	Lib	no	8	OH	19
28	49	49	11-19-'31	Sco	no	7	NJ	20
51	51	56	10-05-'30	Lib	yes	3	NY	21
43	47	71	03-18-'37	Pis	no	5	NJ	22

TABLE 4. (*Continued*)

No.	A Name	B Year(s) elected	C Terms or end	D Party affil.	E Birth state	F Prof- ess'n
23	Benjamin Harrison	1888	1R	Rep.	OH	law
24	Grover Cleveland	1892	1R	Dem.	NJ	law
25	William McKinley	1896, 1900	D10	Rep.	OH	law
26	Theodore Roosevelt	1901, '04	2	Rep.	NY	law
27	William Taft	1908	10	Rep.	OH	law
28	Woodrow Wilson	1912, '16	2	Dem.	VA	law
29	Warren G. Harding	1920	D10	Rep.	OH	jour
30	Calvin Coolidge	1923, '24	2	Rep.	VT	law
31	Herbert Hoover	1928	10	Rep.	IA	engr
32	F. D. Roosevelt	'32,'36,'40,'44	3, D10	Dem.	NY	law
33	Harry S. Truman	1945, '48	p, 1R	Dem.	MO	bus
34	D. D. Eisenhower	1952, '56	2	Rep.	TX	mil
35	John F. Kennedy	1960	D10	Dem.	MA	pol
36	Lyndon B. Johnson	1963, '64	p, 1R	Dem.	TX	pol
37	Richard M. Nixon	1968, '72	1, res	Rep.	CA	law
38	Gerald Ford	1974	p, 0	Rep.	NE	law
39	James Carter	1976	10	Dem.	GA	farm
40	Ronald Reagan	1980, '84	2	Rep.	IL	actor
41	George Bush	1988	1?	Rep.	MA	bus

Explanations:

Column B: Years of election or succeeding to office.

Column C: 2 = completed two full terms; 1R = retired after one term; 1D = defeated for reelection; D10 = died in office; p, R = completed term and retired; p, D = completed term and defeated for reelection.

Column D: Fed. = Federalist; R/D = Republican Democrat.

Column E: State in which president was born.

Column F: mil = military officer; law = lawyer; pol = politician; bus = business man.

TABLE 4. (*Continued*)

G	H	I	J	K	L	M	N	
Age f'st elect.	Age el. presid.	Age at death	Date of birth	Zod. sign	Eldest son	No. of chil.	Death state	No.
24	55	67	08-20-'33	Leo	no	4	IN	23
43	47	71	03-18-'37	Pis	no	5	NJ	24
26	54	78	01-29-'43	Aqu	no	2	NY	25
24	42	60	10-27-'58	Sco	yes	6	NY	26
51	51	72	09-15-'57	Vir	no	3	DC	27
54	56	67	12-28-'56	Cap	yes	3	DC	28
33	51	72	11-02-'65	Sco	yes	1	CA	29
27	51	60	07-04-'72	Can	yes	2	MA	30
54	54	90	08-10-'74	Leo	no	2	NY	31
28	51	63	01-30-'82	Aqu	yes	6	GA	32
48	60	88	05-08-'84	Tau	yes	1	MO	33
62	62	78	10-14-'90	Lib	no	2	DC	34
29	43	46	5-29-'17	Gem	no	4	TX	35
29	55	64	08-27-'08	Vir	yes	2	TX	36
33	56	—	01-09-'13	Cap	no	2	—	37
35	61	—	07-14-'13	Can	yes	3	—	38
38	53	—	10-01-'24	Lib	(?)	4	—	39
56	70	—	02-06-'11	Aqu	(?)	4	—	40
39	65	—	06-12-'24	Gem	(?)	5	—	41

Explanations:
Column G: Age at first election to any public office.
Column H: Age at first election to presidency.
Column K: Abbreviation of birth sign.
Column L: Was president the oldest male child of his parents?
Column N: State in which president died.

tive falling billiard ball example and the purely inductive presidential mortality example. Usually our models require a measure of deduction and a measure of induction.

We have seen that we may run into the need for induction even in as simple a problem as the mass/spring system, if we are unsure of how the spring works or if the spring is nonlinear. In that case our model is hybrid, part deductive and part inductive. We use Newton's laws to the hilt and mix in the results of our experimental measurements, inductively derived information, only when we need to deal with the nonlinear spring constant K. And in using inductivity we must accept the possibility that the predictions using our model will be less reliable than if we had been able to use a purely deductive model. This applies not only to systems such as the mass/spring system, but also subsystems and sub-subsystems.

The important point is this: In designing or formulating a model of a system, the more that we can rely on deduction the better. Each link of the causal chain that is derived from basic laws represents the solution of an analysis problem—one with a unique solution. Each link that is based on induction is one of many possible solutions of a synthesis problem. Each inductive link reduces the probability that the model correctly represents the system and that its prediction will be accurate. We will use the term *predictive validity* or *validity* for short to describe the ability of a model to predict input/output relationships other than those that were used in constructing the model. In these terms, the relation between the amount of inductivity in a model and the validity of predictions can be illustrated as in Figure 11, where increasing darkness is meant to suggest decreasing validity.

THE SPECTRUM OF MODELS

Constituting, as they do, the basis of most intellectual activities, models[2] have many uses and applications. In this book we are particularly interested in using scientific models to predict

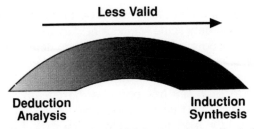

Less Valid

Deduction　　　　　　　　　**Induction**
Analysis　　　　　　　　　　**Synthesis**

FIGURE 11. The greater the amount of induction, the smaller is the predictive validity of a model.

future events. We are, therefore, concerned with how successful these models have been in the past and how much reliance we can place on their predictions. We want to know their predictive validity.

When we examine the many scientific disciplines and subdisciplines that make use of models and simulation, we find that in each application area there are good models and mediocre models, highly sophisticated and elaborate models, and quick-and-dirty models. But let's focus our attention on the best models that each area has to offer, those that are regarded as being most successful and that are most admired. In that case we observe that all the best models of a given area have approximately the same predictive validity. For example, all the best models used to characterize mechanical systems, such as the mass/spring system, generate very accurate predictions most of the time; they have a very high predictive validity. By contrast, models such as that constructed to forecast the mortality of presidents have a very low validity, comparatively speaking. These are extreme examples. The best models of most other areas in the physical, life, and social sciences fall somewhere in between.

In terms of the discussion in the preceding section, we can associate with each application area a characteristic mix of deductive and inductive modeling. In some scientific disciplines, dependable laws, such as Newton's laws, facilitate deduction; in other disciplines, only some vague principles are available; and in some, there are no guidelines at all. Where deductive methods give out, the scientist must fashion the causal links of the model by induction and accept reduced validity as a penalty.

The term *black box* has been used from time to time to characterize systems that must be modeled completely inductively, because the modeler is totally in the dark. In light of the above discussion, we might stretch the "black box" metaphor to encompass boxes of various shades of gray, where the shade of gray corresponds to the approximate relative validity of the model—the lighter the shade of gray, the greater the deductive content of the model and the greater its predictive validity. In this way, we can associate a characteristic shade of gray with each discipline or application area. This is illustrated in Figure 12, which shows a spectrum of models that are used in a number of physical, life, and social science disciplines, all of which make use of models to predict future events. Let's traverse the spectrum from left to right.

Near the "white box" end of the spectrum we find the models that arise in the study of mechanical systems, such as the mass/spring system, and also of electrical circuits. Using Newton's or Kirchhoff's laws and the like, we can construct models virtually without recourse to experiments, to inductivity. We know the laws, we are given the structure, and usually we are also given the magnitudes of the parameters, such as K and M. To be sure, on occasion one parameter or another remains to be found experimentally, but this usually constitutes only a relatively small and uninfluential link in the causal chain.

Proceeding along the spectrum, we next encounter the models of aerodynamics. Here most of the model is well known from basic physical principles, and the structure and dimensions of the plane are also known. However, the lift force acting on an airfoil (like the wing of a plane) is a nonlinear function of the speed of the plane and of the angle that the airfoil makes with the air stream. Another nonlinear function relates the drag force, the air

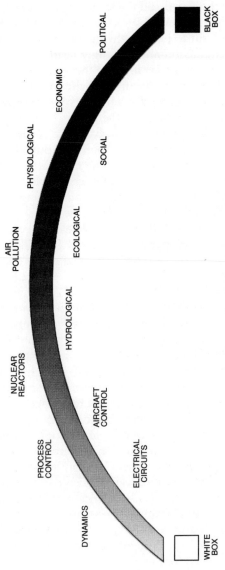

FIGURE 12. The spectrum of models.

resistance, to the speed and the same angle. These two forces are governed by parameters that cannot be derived from theory. We have no choice but to place the wing into a wind tunnel and run experiments. Induction raises its ugly head.

Moving into the grayer areas of the spectrum we come to the models that characterize so-called environmental systems, those that deal with underground water reservoirs, air and water pollution, the climate, and the life. Where there is general understanding of the physical processes involved (the Navier–Stokes equations serve as laws governing the motion of fluids) and of most of the chemical processes as well. But the media in which these processes take place are not readily accessible. As a result, we have great difficulty in specifying or determining all the parameters and functions that have to go into the model. For example, in modeling urban air pollution the wind velocity everywhere in the atmosphere is a key parameter, one that is likely to be very different in different parts of the atmosphere, and one that must be measured. About all we can do is to release balloons here and there and track them as they rise. This makes for a very incomplete and imprecise causal link in the model; induction again.

Further, in the direction of darkness, a variety of life science models are encountered—models of living organisms or parts of organisms. Here there is only an approximate, speculative understanding of the physical and chemical laws, which may underlie physiological phenomena. Furthermore, the structure as well as the parameters of the system are apt to change in an unpredictable manner. When we enter the region of the spectrum that includes psychological as well as physiological aspects, we are even worse off. Now we are no longer certain that there are any laws. Each experimental subject is likely to be unique in some ways, making generalization very difficult and hazardous.

Finally, reaching the social science part of the spectrum, the very concept of "laws" changes. Economists speak of laws such as "the law of supply and demand" and Gresham's Law, but these are really only general guidelines and not at all like the laws that provide foundations for models in other parts of the spectrum. Of course, that does not mean that the models near the very dark end of the spectrum are meaningless or useless. As already pointed out, most of our everyday decisions are based purely on inductive reasoning and on very inconsistent observations. We cannot make decisions without models, and so we do the best we can with whatever models, mathematical or not, that are available to us. Though, we must be very much aware of the great difference in the validities of models falling into different regions of the spectrum. Here is an example.

Example: Take three moments of time in the future. Say, high noon exactly one week from today, exactly one year from today, and exactly five years from today. Suppose we want to predict what each of the following variables will be at each of these three points in time:

1. The distance from earth of the Voyager 1 spacecraft, which was launched in August 1977 and has been cruising through space ever since.
2. The air temperature in Washington, D.C.
3. The state of the stock market as measured by the Dow Jones Industrial Index.

Assume that the best available models are to be used for each prediction. That probably means different models for each of the three points in time. Define "absolute error" as the difference between what the models predict and what actually turns out to be the case. Now compute the "relative error" by dividing the absolute error for each of the three variables—distance, temperature, and dollars—by the expected or observed range of the variables over the five-year period. What we are saying when we compare the validities of the predictions is that the relative error in predicting the location of Voyager 1 will very likely be much smaller than the relative error in predicting the temperature in Washington. And the relative error in predicting the Washington temperature will very probably be much smaller than the error in predicting the Dow Jones average.

Figure 12 is presented to highlight the differences between models as they are used in different application areas, how different the meaning of the term "model" as it is used by different specialists. The spectrum is, of course, far from complete. Some readers may find it an amusing exercise to locate a number of other application areas in the spectrum. But one should not dwell on the spectrum too long or too intently. There are overlaps in the shades of gray of adjacent fields, and there are occasional models that defy classification or that prove to be exceptions. We might say that Figure 12 pictures a model of models, and that model itself is in the dark region of the spectrum.

ARE MODELS GETTING BETTER?

Every scientist working with a model would like to see his model move further to the left of the spectrum, toward the "white box" end, to become more valid. There is the hope and the expectation in the heart of almost every scientist that this will happen some day. If only he could secure more funds, hire a few more PhDs, get more supercomputer time—if only he could get these things and perhaps a little bit of luck, the shade of gray appropriate to his model would lighten, and his predictions of the future would become more valid. How realistic is that hope?

Each of the disciplines in Figure 12 and most others emerged from a prehistory that was unscientific or pseudoscientific at best. Initiates consulted secret codes, astrologers peered at their horoscopes, magicians and alchemists did their best, but the validity of their predictions remained very low. Then, in each discipline, a genius came along—a Newton, Maxwell, Fourier, Marx, Freud, or Einstein—who discovered the key generalizations that eventually became the foundation for the discipline. Often he had charisma as well as ingenuity and quickly acquired a following of students and colleagues. There then began a heady period during which better and better models were introduced, and the validity of these models seemed to improve without bounds. However, eventually in each field the models reached a plateau, a limit of the predictive validity of the models. In some disciplines it took hundreds of years to reach that limit, in others only a few years, but a ceiling was inevitably reached. This is illustrated in Figure 13, which is also a very dark gray model.

After reaching the upper knee of the curve, efforts to improve the validity of the models became progressively less rewarding, and most refinements effected only second-order improvements. It would appear that the ultimate validity to be expected of models is bounded by attributes that are intrinsic to each field: the proportion of deduction and induction that enters the modeling effort. Better data and better experimental equipment may help, but only a little. And only very, very rarely are new basic principles discovered in a mature application area. Most disciplines are therefore doomed to remain anchored to that location in the spectrum that is represented by the asymptote in Figure 13.

Rationality versus Irrationality. We have distinguished between two major approaches of forecasting: unscientific prediction and scientific prediction. Unscientific predictions are generally based on faith—faith in God, faith in the immovability of fate, and faith in the ability of the prophet or seer to discern the future. Scientific

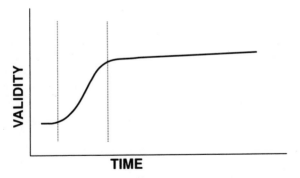

FIGURE 13. Evolution of model validity.

predictions are also founded on faith—faith in empiricism and faith in the scientific method. The difference lies in the philosophy and logical structure that sits on top of the assumptions made on faith. For the scientist, the basic assumptions are remote; he is usually unaware of them in his daily activities. Instead, the scientist places great emphasis on the logical structure and the logical inferences that permit far-reaching conclusions to be drawn from observations and experiments. The scientist takes pride in being rational. In his view, science is based on *rationality*, while unscientific approaches to prediction are irrational. Nonscientists may disagree with this dichotomy, but in most of our western culture a scientist commands respect and acceptance because he is a rational animal, because he subscribes to the scientific method, and because he validates his conclusions using logical proofs and controlled experiments. In the following chapter we will take a closer look at the prediction of catastrophes and the problems imposed on the scientist when controlled experiments and validations are not possible.

magnitudes, would be enormously time consuming for even the most talented of us working alone. And that is where computers come in.

Using clever software and programming techniques, scientists implement the mathematical model on the computer. To that end, they translate the mathematical model into a computer language, such as FORTRAN. Each equation and initial condition is represented by one or more statements in that language. They either have to learn how to do this or hire someone who does. They then enter the program into the computer, usually by typing at the keyboard. Next they type in the numerical values that they want to use (for example, D, M, and K of the mass/spring system). After a while they receive the outputs that they request, perhaps as tables of numbers, perhaps as graphs, or perhaps as colorful animated "visualizations."

It is feasible to type equations directly into the computer and to ask for a solution in analytical form, such as Eq. (2) in Chapter 3. If desired, numerical values for the parameters can be inserted later. That is called *symbolic manipulation*, and some powerful software is available to help. It is also possible to frame models as logical statements, such as the presidential mortality model, and avoid the use of mathematics. But for large, complicated models, these methods use up too much computer time and memory. For this reason, most of the action is in *numerical computing*.

Scientists start with a mathematical model and employ a technique called *numerical analysis* to transform the model into another set of equations that their computer can handle more conveniently. Computers are really quite limited in the kind of mathematics that they can do. They can add, subtract, multiply, and divide, but they cannot do trigonometry or calculus without making approximations. In the transformed model, the magnitudes of all parameters and initial conditions appear as numbers—the values that the scientists choose to assign to them. All inputs are likewise included as numbers or series of numbers. If an input is a transient, it is represented by entering numbers corresponding to the input at successive instants of time (say every second).

These numbers are called "samples" or the time-varying input variable.

The computer then races through a long series of arithmetic operations, in accordance with the program that has been prepared and entered into the computer. Eventually, the computer generates the required output—usually also as a series of numbers, samples of the output transients at successive instants of time. The whole process may take a lot of arithmetic operations and is therefore called *number crunching*. As the model grows in complexity, the number of operations ($+$, $-$, $\times$, and $\div$) that have to be carried out grows astronomically and calls for more and more powerful computers. The best of our contemporary computers can easily perform thousands of millions of operations per second.

Today's computers are devoted to two major classes of tasks: business data processing and scientific computing. In terms of hardware cost, business computers take perhaps 60% of the market and scientific computers about 40%. And most of the scientific computations involve models and the solution of the equations constituting the models.

Ever since the end of World War II, scientists and the manufacturers of scientific computers have enjoyed a close symbiotic relationship. The appetite of scientists for computers to implement ever larger models of physical systems—weather forecasts, aerospace designs, weapons of all kinds, among others—encouraged the computer industry to come up with generation upon generation of supercomputers and also encouraged the U.S. government to find the money to pay for them. The term *supercomputer* is used to characterize the fastest and most expensive scientific computer *currently* available. The term therefore describes a moving target. As mentioned before, today's workstations and large personal computers selling for under $20,000 are every bit as powerful as the then world-class supercomputers that cost several million dollars in 1978, while today's supercomputers are millions of times as powerful. And the end is by no means in sight.

The term *computer simulation* is used to describe the implementation of models on computers in such a way that the user can explore how the changing of parameters, inputs, and even the structure of the model affects the output. Computer simulations permit the scientist to run "experiments" on the computer. These computer experiments presumably provide the same outputs that would be observed in the real-world system represented by the model, if it were possible or practical to experiment with the real-world system. It is not unusual to generate hundreds or even thousands of sets of computer outputs, each set resulting from a different combination of parameters and inputs. The entire simulation approach is illustrated in Figure 14.

The overall methodology for simulating systems in physics, chemistry, physiology, psychology, or economics is pretty much the same. In one way or another it involves each of the following steps.

Problem Formulation. The scientist prepares a lucid and concise description of the information that is desired. This is the statement of objectives and must precede any effort to model the system.

Mathematical Modeling. A judicious mix of induction and deduction is used to provide a set of mathematical expressions that permit the calculation of the output.

Numerical Analysis. Standard mathematical techniques are used to translate the mathematical model into another model, one that the computer can handle. This is an algebraic model that can be solved without needing to perform calculus.

Implementation. The transformed model is expressed in a language that the computer understands. A computer program is prepared. This program is entered into the computer (often by typing at a computer terminal). There follows an often frustrating and time-consuming series of tasks to "verify" that the program in the computer actually corresponds to the mathematical model, that no errors have been made in numerical analysis or in programming. In large simulations this debugging phase may take

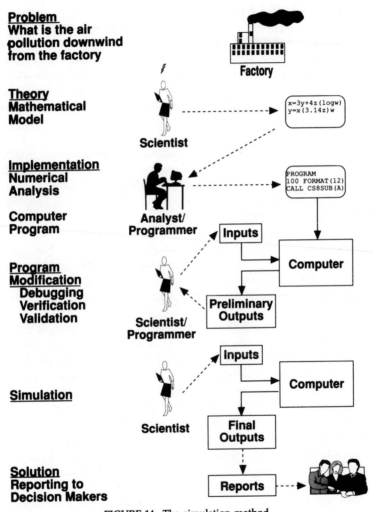

Problem
What is the air
pollution downwind
from the factory

Factory

Theory
Mathematical
Model

$x=3y+4z(\log w)$
$y=x(3.14z)w$

Scientist

Implementation
Numerical
Analysis

Computer
Program

Analyst/
Programmer

PROGRAM
100 FORMAT(12)
CALL CS8SUB(A)

Inputs

Computer

**Program
Modification**
Debugging
Verification
Validation

Scientist/
Programmer

Preliminary
Outputs

Simulation

Scientist

Inputs

Computer

Final
Outputs

Solution
Reporting to
Decision Makers

Reports

FIGURE 14. The simulation method.

weeks or months, and even then there is no absolute certainty that all sources of error have been eliminated.

Validation. Next it is necessary to find out whether the model on the computer, now presumably nearly free of all bugs, is a sufficiently correct representation of the system being modeled—whether and to what extent it is valid. This often entails an exhaustive and exhausting series of computer runs (more on this later).

Simulation Runs. After all this work, the simulation is finally ready to provide useful and interesting outputs. And to justify all the time and expense, hundreds or perhaps thousands of computer runs are made, each generating a different set of outputs. This is discussed in the next section.

Utilization of the Results. The outputs and all that has been learned from them are compiled in detailed reports and in summaries. Eventually, these are communicated to the public and to interested decision makers in government, industry, and academia.

That is the simulation methodology—the simulation paradigm. The process may take a day or less for a simple "white box" problem or decades for something as complex as global climate. But it is a methodology that has proven itself time and again. Many of our modern technological achievements would have been impossible without sophisticated computer simulations: modern jet aircraft, the conquest of space, nuclear and alternative generators of energy, satellite communication, and solid state integrated circuits to name just a few. Simulation is also becoming increasingly important in medical diagnosis and research, and many economists would be lost without their computer models. There have been disappointments and failures along the way, of course. But, with bigger and better computers arriving in a continuous stream, the simulation Community is bright eyed and bushy tailed.

MANIPULATING THE SIMULATION

A scientist who succeeds in getting a simulation to run on a computer and who has gained confidence that it is running properly is like a child with a new toy. Unlike a scale model or a laboratory setup, a simulation is infinitely flexible. A few simple commands entered from a terminal can effect profound changes in the model and the way it transforms inputs into outputs. Here is the opportunity to ask: "What would happen if. . . ?" and to get the answer in short order. "Suppose the sun rose in the west instead of the east. . . ," "Suppose the gravity on earth were that of Jupiter . . ."; for a while, the scientist feels like a master of the universe.

But then the serious work begins—tuning and improving the model, validating the model and the outputs, and establishing the credibility of the model. This is invariably a slow and arduous experience—one that is the real test of the scientist's mettle and dedication.

To illustrate this process, let's return to the simple mass/spring system discussed in the preceding chapter and reproduced in Figure 15. The equation constituting the mathematical model contains the mass M and the spring constant K. These are the two

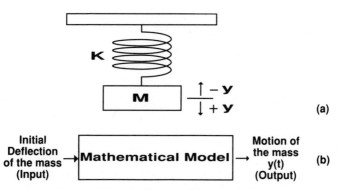

FIGURE 15. Schematic diagram of the mass/spring system.

parameters of the model and appear as coefficients in the equation. For each simulation run, specific values for these parameters are selected and entered into the computer. The output is the deflection, y, plotted as a function of time. In this case it looks like a wave of constant amplitude and frequency. In order to explore the behavior of the mass/spring system, the scientist might try out say a dozen different combinations of M and K and generate an entire family of curves as his output.

As a next step, the scientist might reason that the description of the property of the spring as a constant, K, is too simple-minded. Actually, as some springs are stretched more and more, they become weaker and weaker; it takes less and less additional force to stretch them even farther. In other words, as y increases, the spring constant, K, decreases. Accordingly, the scientist might modify his model and computer program so as to represent K as a function, a curve as shown in Figure 16. K is thereby specified to be a *nonlinear function* of y. Possibly this curve can be obtained from a handbook; or perhaps data from laboratory experiments with an actual spring are available. Either way, sampled values of the curve are stored in the memory of the computer and "called" for when needed.

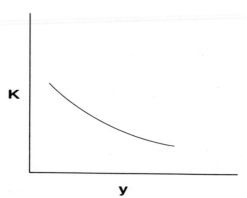

FIGURE 16. Example of a nonlinear spring characteristic.

This means that as the computer goes through its paces, as it calculates the displacement y at successive instants of time, it must refer repeatedly to a table in its memory. That table contains the information provided by Figure 16, and may mean, for example, that when $y = 2.5$, $K = 3.7$; but when $y = 4.5$, $K = 3.2$. This increases the time to get a solution, but it opens up a whole new universe of possibilities. The scientist can now investigate the effect on the output of using springs with all kinds of non-linearities, all kinds of curves, like the family of curves shown in Figure 17. Of course, the scientist can also study the effect on the output not only of constant inputs (a simple displacement D of the mass), but also of input transients.

Eventually the choice of combinations of parameter, functions, and inputs becomes too vast to permit an exhaustive study of all possibilities. Instead, the scientist selects what he considers to be an interesting subset of the various possibilities and generates outputs only for the members of this subset. The term *scenario* is often used to describe a specific combination of inputs and system properties, selected by the scientist, and the resulting outputs. Hundreds or thousands of such scenarios may be run in the course of a challenging simulation. How does the scientist

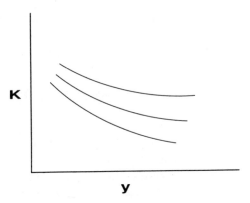

FIGURE 17. A family of curves.

decide which scenario to use? How does he decide when he has enough information? It all depends on the objectives of the simulation.

If the objective of the simulation is to predict future events so that appropriate preventative or mitigating actions can be taken, the scientist must be concerned with the credibility as well as the validity of his results. He must answer the following questions for himself.

1. Am I confident that the model as implemented on the computer constitutes a sufficiently correct and accurate representation of the system or phenomenon being modeled? How valid is the simulation?
2. Are the results and accompanying documentation sufficiently coherent and incisive to be accepted by decision makers in particular and the public in general? Should I go public with the results now, in the face of possible disagreements and controversy?

As we will see in the following sections, the scientist is on the horns of a dilemma. On one hand, the job of validation is never truly complete. There always remain uncertainty, insecurity, and gnawing doubts, all impelling the scientist to make yet another series of computer runs, yet another modification of the model. On the other hand, computer runs cost money, often lots and lots of money, and there are many pressures to meet deadlines, to finish up and to publish. Whether he likes it or not, the scientist is compelled to consider all sorts of trade-offs and compromises.

WHY DO SIMULATIONS GO ASTRAY?

Even with the most powerful of computers and the most skillful and dedicated of scientists, many simulations yield incorrect or misleading solutions. The general reason for this is simple: the outputs of simulations are invalid whenever the model constitutes an incomplete and/or inaccurate representation of those attributes of the real-world system that affect the outputs. But now we have to look at the words "incomplete" and "inaccurate" a bit more closely. We will do this by exploring some of the basic assumptions inherent in the modeling methodology, first in regard to the inputs and then in regard to the system.

Separability. To some extent all phenomena in the universe are interrelated. A gust of wind in India blows a girl's hat off her head, and four years later in Paris a young man proposes marriage to her. An imaginative novelist can provide a chain of circumstances linking the two events. Or, if you were to pick up this book and strike it sharply against the arm of your chair, you would initiate molecular motions, most very small to be sure, and nothing in the world would ever again be precisely the same.

So, strictly speaking, every system is subjected to inputs that are the result of the interactions of many, many processes through the world. In modeling a system, however, we assume that almost all of these interactions have a negligible effect on the system outputs, that they can be ignored. This permits us to define the system as a separate entity. We first circumscribe the system by specifying its boundaries or by enumerating its components. In doing so, we make clear what belongs to the system and what belongs to the rest of the world. And we assume that the only way that the rest of the world can affect the outputs of the system is by way of the inputs that we specifically describe. That is the assumption of *separability*, and it is fraught with disaster.

Consider the mass/spring system of Figure 15. In the model that we have presented, we have assumed up to now, implicitly, that the only input that affects the way the mass bobs up and down is the initial displacement of the mass. But can we be sure that that assumption is justified in a real-world system? Perhaps the wooden beam to which the spring is attached is not perfectly still. Perhaps it is actually in motion, vibrating because a heavy truck is passing nearby. Or perhaps the whole system is actually mounted in a truck and is being bounced around as the truck negotiates a rough road. That would certainly affect the motion of the mass, and we would find its actual movements to be quite different from those predicted by the simulation. The simulation would then be invalid, and we would have to go back to the drawing board and come up with a better model, one that includes the motion of the truck as an additional input.

In the case of the mass/spring system we can harbor the hope that we can eventually discern and include all relevant inputs. However, when we model something as complex as the weather or the economy, we can never, never be sure that we have not omitted inputs that may under certain circumstances have a very significant effect on the outputs. We just have to live with that uncertainty.

Selectivity. In all real-world systems a wide variety of phenomena coexist—many physical processes, chemical processes, biological processes, economic processes, etc. When we model a system, we usually focus our attention on one or a small number of these processes and assume that all of the others have a negligible effect upon the outputs of interest. That assumption may or may not be justified.

In modeling the mass/spring system, we ignored all physical phenomena except for the simple mechanics governed by Newton's Laws, and we hoped for the best. It so happens, however, that spring constants of metal springs depend upon temperature. The warmer the spring, the smaller K. Perhaps we should have included the air temperature in our model. Furthermore, when a spring is stretched and compressed many times in succession, as would be the case if the mass keeps bobbing up and down for a long time, the friction of the metal particles in the spring as they rub against each other generates heat. Hence, the longer the mass vibrates, the warmer the spring and the smaller is K. Should these effects be included in the model? It all depends on by just how much K varies and by how much that affects the output. More uncertainty as to the validity of the model.

Chaining of Model. Often in the modeling of complex systems a number of models are connected end-to-end to form a chain. For

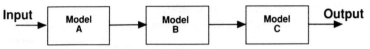

FIGURE 18. A chain of three models.

example, in Figure 18 we see a model that consists of three submodels, Models A, B, and C. The output of Model A serves as the input of Model B, and the output of Model B is used as the input of Model C.

In environmental simulations, Model A might describe the way a pollutant is emitted by various factories, power plants, automobiles, and other sources of pollutants. The output of Model A is then the total amount of some chemical, such as sulfur dioxide, that winds up in the atmosphere. Model B might then use this information to predict the eventual precipitation of sulfur compounds on land and water surfaces. The output of Model B is then the geographic distribution of the chemicals that pollute our food and drinking water. Model C employs the output of Model B to predict the effect of these pollutants on human and animal health. Thus we start with a knowledge of the chemicals emitted by the various human activities, and we wind up with predictions of increased rates of cancer and other maladies.

When this chain is used to predict the "output" on the basis of a specified "input," the errors inherent in each of the three models contribute to the error in the output, the error in the predictions generated by the simulation. Suppose, for instance, that in terms of the notion of "relative validity," introduced in Chapter 3, Model A has a validity of 0.7, Model B has a validity of 0.1, and Model C has a validity of 0.8. The validity of the chain consisting of the three models would be

$$0.7 \times 0.1 \times 0.8 = 0.056 \text{ or } 5.6\%$$

In other words, when we put models end-to-end, the errors introduced by the first model are compounded by the errors introduced by the second model, and so on. The final output is far less reliable, far less faithful to reality than any of the component models. We have a chain that is far weaker than the weakest link.

The Curse of Dimensionality. The models of most systems are complex chains (or networks) of subsystems linking the inputs and outputs. But if we look closely at any of the links of such a chain, we can usually discern a series of smaller links, and under a magnifying glass each of these links is revealed to consist of even smaller links. For example, the spring of the mass/spring system was represented as a single element or component with spring constant K. But we know that a spring is a helix formed by a series of loops, each of which itself acts like a spring. Therefore, we might represent the spring as, say, ten smaller springs connected end-to-end. And if we look at each loop under a microscope, we see a larger number of metallic crystals; perhaps these should be shown in the model. How far do we go? It is up to the modeler to decide at what point to stop this disaggregation and to accept an admittedly coarse approximation of the fine structure of the elements of the system.

Disaggregation is not just a matter of size. Quality is involved as well. In addition to mass and spring constant, the system surely has other physical properties. For example, because of the friction within the spring, the energy imparted to the mass/spring system when the mass is initially displaced is gradually used up or dissipated. As a result, each time the mass bobs up and down its maximum displacement becomes a little smaller than it was during the previous excursion. Eventually, the amplitude of the wave goes to zero, and the mass comes to rest. In order to represent this effect it is necessary to make the model more complex.

There are many other physical phenomena that could be encompassed by the model as well, phenomena that under certain circumstances could have important effects on the output. How many of these phenomena should we include?

The problem is that every time that we carry the disaggregation a step further and every time we add a new phenomenon to our model, the outputs become more difficult and more time-consuming to compute. We are adding a dimension to the computational complexity each time we do that. Including the damping constant R in the model of the mass/spring system increases the number of parameters from two to three. This makes the system many times more difficult to simulate. Now to study the system completely, we have to consider all possible combinations of M, K, and R, instead of just M and K. If we are satisfied with only ten possible values of each parameter, that means that we now have to make $10 \times 10 \times 10$ or 1000 computer runs, instead of only 100. And each additional parameter or dimension that we add to our model will multiply the total number of runs by another factor of 10.

When we try to simulate much more complicated systems than the mass/spring system, the number of parameters and nonlinear functions becomes very large. This makes it virtually impossible to make a thorough study of all likely possibilities. The late mathematician Richard Bellman used the expression "the curse of dimensionality" to describe his frustration with his inability to grapple with this explosive growth of the number of dimensions of system models and with the consequent disastrous increase in computational complexity.

Inevitably, therefore, in the simulation of complex systems, the scientist is compelled to ignore some of the parameters or dimensions of the systems, dimensions that he suspects may be significant. And in doing so he risks diminishing the validity of the simulation. Each reduction in dimensionality increases the chances of misleading interpretations of the outputs. The situation is similar, in some aspects, to trying to identify a man by the shadow that he casts on the ground. Easy enough to tell that it is a person, but it is unlikely that enough details would be available to identify the individual. In the case of the mass/spring system, there is sufficient comprehension of the physical phenomena involved so that a scientist can be reasonably confident that his model has a dimensionality adequate to satisfy the objectives of the simulation. That is rarely the case for more complicated systems.

The curse of dimensionality becomes especially vexing when dealing with models of systems in the dark regions of the spectrum. There then exist few guidelines as to which of the multitude of parameters should be included in the model. The presidential mortality example discussed in Chapter 3 is a case in point. There have been forty-one presidents in all, and an unlimited number of attributes could be associated with each of them. Lacking a plausible causal chain, we have no way of knowing how extensive a search of these parameters we should make before accepting that the "election in a year that is a multiple of 20" parameter is sufficiently significant to be used to predict the fortune of future presidents.

Going Out-of-Bounds. In the case of the mass/spring system, the larger that we make the initial displacement of the mass, the larger the amplitude of the resulting vibrations. In fact, if we neglect damping, the lowest point that the mass will be predicted to reach during each cycle of bobbing up and down will be the same as the point to which it was initially displaced. And the highest point that the mass would be predicted to reach during each cycle will be exactly as far above the equilibrium point (location when the mass is at rest and where the displacement, y, is zero) as the mass was initially displaced below the equilibrium point. In the simple model of Figure 15, if we double the initial displacement, D, we also double the displacement experienced by the mass at every instant of time thereafter. We have already discussed the concept of nonlinearity, the possibility that K may not remain constant for all displacements (see Figure 16). In other words, doubling the input does not necessarily double the output. Usually, the approximation of linearity, that K remains constant, is close enough to the truth, but only up to a point.

If in experiments with an actual spring, we keep increasing the initial displacement, D, initiating vibrations of larger and larger amplitudes, we eventually reach a point where the simple model breaks down completely. In the upward motion of the mass, the loops of the spring are then forced more and more closely together; and for very large motions they will be compressed against each other, blocking any further upward motion. This has the effect of making K very large. Also, if we initially pull the mass too far down, if we stretch the spring too much, we may permanently alter the spring so as to make K much smaller than it was before. This is shown in Figure 19.

In other words, if we make the initial displacement too large, the model of Figure 15 is no longer valid; it is no good for predicting the motion of the mass. We would then need a new model, including the information in Figure 19. So it is essential to keep in mind that the model of Figure 15 is based on the assumption that we will not try to make the excursions of the mass exceed a definite limit—that we will not go out-of-bounds.

Unfortunately, unless we have run extensive tests on springs and observed how they respond to large displacements, we have no way of knowing exactly what those limits or bounds are. And some of the tests we need to perform may well damage the springs. As a result, whenever we make a simulation run in which we cause the displacement that is greater than one that we have

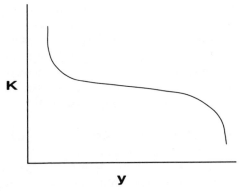

FIGURE 19. Spring characteristic when the spring is stretched beyond its limits.

observed in real life, we run the risk of accidentally going out-of-bounds and getting invalid results.

All system models, without exception, have bounds that limit their applicability. Even the simple "drop a ball from the top of a tower" model is limited. The governing equation does not apply if the tower is very high. If it is so high that it extends beyond the gravitational field of the earth, the ball will not fall down at all, but will wander off into space. In that system we are, however, quite clear about the bounds that limit the validity of the model. This is not so for complex systems.

Let's return to the presidential mortality example. We saw that a model spanning the period from 1860 to 1960 is remarkably consistent—no exceptions. But what about before 1860 and after 1960? We now know that the model failed in 1980, but prior to 1989 we had no way of knowing that it would. Since we have no inkling of a causal chain, we have no way of placing meaningful bounds on the period of time that the model remains valid. Of course, this is a rather extreme example, a very "black box."

But even in the light gray region of the spectrum, we are invariably haunted by the "out-of-bounds" problem whenever we are forced to use inductivity in fashioning some of the links of our causal chain. Induction means observing the system in action, the response of the system to observed or applied inputs. How can we be sure that the behavior of the system would not change abruptly if the inputs were to increase in magnitude beyond the magnitude of the inputs that we have been able to observe? Most of the time we are left with the uncomfortable knowledge that our model will fail to be valid when we try to simulate phenomena outside a range, the limits of which we do not know. And there is still another phenomenon to make our lives difficult.

Chaos. According to Newton's Laws, the direction and speed with which a body, any body in the universe, moves is determined entirely by the forces acting on it. We can calculate these forces, not only for a single body but also for any number of bodies, by solving an analysis problem—a problem with a unique solution.

It follows that if, at some instant of time, we know the locations and velocity of all bodies in the universe, we would theoretically be able to predict the location and velocities of all of the bodies for all future time. In other words, if we could have a complete knowledge of the present, we could predict the future for all time. The future is completely determined. That is the *mechanistic* or *deterministic* worldview.

This mechanistic idea prevailed throughout the nineteenth century. But with the advent of the theories of Albert Einstein and of the coterie of "modern physicists," this worldview had to be revised. When looking at very small particles, such as the components of atoms, scientists found that it is no longer possible to assign a specific location and velocity to each particle. Rather, one deals with probabilities. We can talk about the average position or speed of a group of particles, but not about the trajectory of individual particles. This new nondeterministic approach at the microscopic level made possible the great advances in atomic and molecular physics and chemistry. However, it did not necessitate a revision of the methods used to model systems at the macroscopic level. Newton's Laws and determinism sufficed. They were adequate for the modeling of droplets of water, of balls dropped from towers, of mass/spring systems, and for mapping the trajectory of planets and stars. In all these areas, it was assumed, that if one had the correct inputs and the correct model of the system, then the outputs were uniquely determined and could be computed by simulation.

In the 1970s, this view had to be revised as well. It was discovered, quite by accident, that many nonlinear systems have unpredictable responses to excitations. This discovery has been the subject of intensive study for the past twenty years, much of it using computer simulation. It appears that many systems, even simple systems such as the mass/spring system, are characterized by nonlinear differential equations that have a very strange property. The system outputs are sometimes enormously sensitive to very tiny changes in the magnitude of the initial conditions or inputs. Very small changes in the inputs can cause the outputs to take on a totally different form. For our mass/spring system, that means that if the spring exhibited certain kinds of nonlinearity, a minute change in the initial displacement, D, might result in a major change in the way the mass bobs up and down. It would therefore be impossible to predict with confidence exactly how the mass would move. That would depend upon precisely where we release the mass. This phenomenon has been given the picturesque name *chaos*.

We now know that chaos is pervasive. It is all around us. It has been observed in astronomy, biology, ecology, and even in economics. We are usually safe in ignoring it in most system simulations. The deterministic method remains applicable most of the time, although we must always be on guard in designing our models and interpreting the result of simulations. However, there are a number of application areas in which the discovery of chaos has forced scientists to modify their use of deterministic models and their confidence in the outputs of their simulations. For example, twenty years ago meteorologists were very confident that with bigger and bigger supercomputers it would eventually be possible to make long-range forecasts of the worldwide weather by modeling and simulating the atmosphere as a deterministic system. Now few meteorologists believe that mathematical models will ever permit forecasts farther ahead than a week or two. The reason: chaos. So here we have yet another potentially significant source of error, degrading the validity of simulation models.

THE ROLE OF STATISTICS AND PROBABILITY

As we have seen, the modeling and simulation of real-world systems is rife with uncertainty and doubt. The mathematical discipline of probability and statistics was made to order to deal with situations in which data are not uniform, exhibit ambiguities, or are of questionable value or significance. It is not surprising, therefore, that this discipline plays a vital part in many of the tasks and activities involved in the modeling and simulation of complex systems. The detailed treatment of that subject is beyond the scope of this book. *Forecasting in the Social and Natural Sciences*, edited by K. C. Land and S. H. Schneider,[1] provides a fine overview of the role of statistics and probability in the modeling of many of the systems discussed in the following eight chapters. H. W. Lewis[2] in his recent book, *Technological Risk*, presents a very readable introduction to the pertinent concepts of statistics and probability. He then carefully considers the assessment and the management of the risks associated with industrial chemicals, highway and aircraft transportation, fossil fuels, and radiation. Here we will offer only some brief comments that pertain particularly to the prediction of catastrophes. Recent books by J. Casti[3] and by M. G. Morgan and M. Henrion[4] provide very readable discussions of the use of statistics and probability to look into the future.

Statistical Data Analysis. The raw materials for the modeling of systems or subsystems by induction are the observations or measurements of the system inputs and outputs—of data. By means of the well-established method of *regression analysis*, all these data can be systematically analyzed to determine which input variables appear to have a substantial effect on the outputs. For example, the family histories of a large population of cancer patients can be processed to provide an indication of genetic predispositions to certain forms of the disease. Such indications can serve to alert scientists to the possible existence of causal chains and thereby to better focus their research effort in promising directions. Statistical methods are also widely used to evaluate the significance of observed relationships between inputs and outputs: this to suggest whether apparent correlations are really just a matter of chance, and if not, how much confidence should be placed in the data.

In this way, as well as in a number of others, statistical methods have been invaluable in a host of challenging and important applications, especially in the life and social sciences. However, we must not lose sight of an important fact. Statistics can only serve to enhance the confidence of the scientist in the validity of a model by helping to confirm a suspected or proposed causal relationship. The mere existence of a statistical correlation cannot establish that a causal relationship actually exists. It takes a plausible explanation of the reasons for that relationship to do that.

If a system problem is in the extreme dark end of the spectrum, if there are no accepted plausible theories as to why the observed inputs seem to affect the outputs, statistical studies provide little comfort or confidence in the validity of a model. The presidential mortality example is a case in point. Here there is a very strong correlation between the observed input (year of election) and the observed output (death while in office). In fact, the variance is zero for the period from 1860 to 1960. Yet there is no plausible theory, and common sense tells us that the apparent correlation must be accidental and of no significance. Incidentally, statisticians would hasten to point out that the six elections involved do not constitute a sufficiently large sample to permit definitive conclusions.

Probability. On many occasions, reports of the results of simulations are couched in probabilistic terms. There are many different ways of arriving at these probabilities. Here we mention only three inherently different approaches.

In many simulations the magnitude of some of the parameters or inputs may not be known precisely or may vary from time to time. Under these conditions, parameters or inputs can be expressed as statistical distributions and handled as such in the simulation. For example, in the mass/spring system, the mass M may be known always to lie in the range of 0.7 to 1.3. Suppose moreover that measurements or theory suggest that 60% of the time the mass is in the range 0.9 to 1.1; 25% of the time between 0.8 and 0.9 or between 1.1 and 1.2; and 15% of the time between 0.7 and 0.8 or between 1.2 and 1.3. Large numbers of simulation runs are then made, and a so-called random number generator is used to set the value of M for each run, such that the specified distribution is maintained over say 1000 runs. The results of the individual runs are then combined and averaged. This is called *Monte Carlo* simulation, since the output of each run is governed to some extent by probabilities and chance.

In some simulations in which the model consists of a network of subsystems, a probability may be assigned to each link of the chain. This permits the characterization of the overall system in terms of a probability. For example, a nuclear reactor may contain many components which are potential sources of failure or accident as well as numerous redundant circuits and safety mechanisms. The probability of malfunction of each component can be determined independently by experiment or by theory. These figures can then be combined to predict the overall probability that the reactor will fail.

Neither of these two probabilistic measures relate to the validity of the model of the system. The probabilities as reported are only as valid or invalid as the model used to compute them. However, on occasion, a probability is attached to the validity of the model. When it comes to catastrophes, this is always a subjective measure. Each member of a group of experts is asked to make an estimate of the probability that predictions made on the basis of a model and a simulation are correct. The opinions of the experts are then averaged.

Pitfalls and Popular Misconceptions. The field of probability and statistics has a long and illustrious history. It has a strong formal component that permits experts in the field to communicate with each other with a minimum of misunderstanding. The problem is that it is usually very difficult for them to communicate the details and the implications of these formalisms to the users of the statistical models. As a result, decision makers and the public are frequently mystified and misled.

Consider, for example, a radio announcement stating: "The probability of rain in New York on March 15 is 50%." What is the 50% figure supposed to mean? If the announcement is made in July, it probably means that somebody has examined the weather reports for March 15 for the past 50 or 100 years and found that rain was reported in one half of those years. If the announcement is made in February, the 50% figure might be based on a study, perhaps a simulation, of the general weather patterns for that winter. If the announcement is made on March 13, it may mean that in the recent experience of a meteorologist, the weather pattern observed on that day has led to rain two days later 50% of the time. And that figure may have been based on extensive computer runs or it may reflect the gut feeling of the weather forecaster. The public is rarely given enough information to know which of these applies. Also, New York is spread out over a very large area. Does the 50% probability refer to rain everywhere in the area, or only to a specific location within New York? And how do you distinguish "rain" from a "drizzle" or "heavy fog"? There are good answers to all of these questions, but these answers are usually concealed from the public and even from sophisticated users of the information.

And then there are fundamental semantic problems. Not long ago, a famous sportscaster was heard to say: "Well, well, well, the weather forecast for today was for a 20% probability of rain, and now the birds are singing, the sun is shining bright, and it's a great day for a ball game. So the weatherman was wrong again." This statement may have caused a few science cognoscenti to chuckle wryly or to grimace in anguish. But it undoubtedly was absorbed uncritically by the overwhelming majority of the listening audience, totally oblivious to two serious insults to the theory of probability.

First, a prediction couched in probabilistic terms cannot be considered to be "right" or "wrong" on the basis of a single observation. The weatherman is subject to criticism only if he makes a "probability of 20%" prediction for say 100 different days, and it turns out that substantially more or less than 20% of those days happen to be rainy.

Second, when the forecast specifies a "20% probability of rain," the weatherman is saying, in effect, "The chances are that it will not rain. In fact, I would bet four dollars to your one that tomorrow will be a fine day." Yet many people hearing this prediction will be led to believe that they had better look for their raincoat and umbrella and forget about going to the ball game. The source of the misunderstanding lies not so much in our deficient educational system but in the way that we use the English language, and in fact most modern languages. When we say: "It will probably rain," we mean: "It's likely to rain; not for sure, but very likely." Unfortunately, many people wrongly believe that the phrase "a probability of $X\%$" is synonymous with "probably," regardless of the number that is represented by X, whereas to the weatherman that number makes all the difference. Along the same lines, most people consider the statement: "There is a probability of rain" to by synonymous with: "It will probably rain," when in fact saying: "There is a probability of rain tomorrow" is like saying: "There will be weather tomorrow"—true, but totally uninformative.

The point is that we are naturally led into all kinds of ambiguities and miscommunications when we attempt to introduce probabilistic concepts into our everyday discourse. This, unfortunately, applies to many of the results of simulations that are reported as probabilities and especially those that involve catastrophic events. It is difficult enough for specialists to use the concepts of probabilities with consistency and precision in talking with each other. When scientists use probabilities in communicating to the general public, they often end up being misunderstood.

VALIDATION AND CREDIBILITY

In this chapter we have examined some of the major pitfalls inherent in the modeling and simulation of complex systems. Is it really as bad as all that? How can we ever believe what the computer tells us? A scientist must grapple with those questions all the time. And he is faced with two distinct issues:

1. How confident am I of the results of my simulation?
2. How am I to convince decision makers and the public to take the results seriously?

The first of these relates to the validity of the model, and the second relates to its credibility.

Validation. Validation is the never-ending task of proving that the simulation model represents the system with sufficient accuracy and in sufficient detail to satisfy the objectives of the simulation. There is no cut-and-dried way of going about this. Every scientist charts his own course; here are some of the most widely used approaches that he may employ to increase his confidence in the model.

- Examine and reexamine the model to confirm that the causal chains that it embodies are indeed plausible and have remained plausible, even in the light of newly emerging or evolving theories.
- Examine the final as well as intermediate simulation outputs to determine whether they are plausible in the light of insight and knowledge of the system and phenomena being modeled (e.g., if an airplane is predicted to fly at a negative elevation, something is very likely wrong with the model).
- Compare the simulation outputs with the outputs of other simulations of the same system and try to reconcile any differences.
- Check the ability of the simulation to correctly predict outputs caused by inputs other than those used in constructing the model. The scientist may use forethought to this end and "save" some system observations in order to use them later for validation.
- Check the prediction of future events as generated by the simulation against the corresponding real-world events as these occur.

The conscientious scientist will do all of these things, and modify his model again and again, until he finally believes in his results or until he collapses from exhaustion. Inevitably in the process, he continuously learns more about the system and the phenomena being simulated. He gradually develops a profound and intimate feel for the system—its properties and its peculiarities. Sometimes all this increases his confidence in the simulation above and beyond objectivity. The simulation is his creation, his baby, and it has his full trust and faith.

Usually, though, some nagging doubts remain. These doubts increase, sometimes unbearably so, as the moment of truth approaches—the day that the simulation results are to be revealed to the outside and sometimes hostile world. I vividly remember waking up in the middle of the night, not once, but on a number of occasions with various thoughts of doom in my head—I was suddenly convinced that my model was worthless, that I had forgotten some essential and fatal impediment, that I would become a laughing stock, etc. Most scientists are similarly diffident about releasing their results, and they repeatedly try to postpone the day of reckoning. But in doing so, they face mounting pressures.

The Pressure to Publish. Simulation is big business. A state-of-the-art supercomputer can cost in excess of twenty million dollars to buy and millions per year to maintain and operate. Every microsecond of computer time is therefore precious, and somebody has to pay for it. Unfortunately, very few scientists are independently wealthy and therefore able to call their own shots.

If the scientist is in industry or works for a government laboratory, his supervisors and managers have had to push hard to secure the necessary financial and technical resources. The powers that be expect results promptly. If he is in academia, the funds will probably have come from a government agency in the form of a grant or contract. No matter where he works, though, he will have had to submit proposals in competition with other researchers, proposals that highlight the importance and timeliness of the expected results, and he will have committed himself to various deadlines. He will have to submit frequent reports and be subjected to periodic visits from the project managers of the funding agencies, not to mention his own superiors. And everybody wants to see results, results, results—or no more money.

If the scientist is a junior member of a university faculty, he lives in a "publish or perish" environment: no results, no technical papers published in prestigious journals; no papers, no tenure. If the scientist is still a graduate student, he may enjoy the shelter provided by the ivory tower, but only for a while. Eventually, the real world will catch up with him too. And even the most prover-bially patient of spouses have been known to fling soggy diapers and shout: "Finish that # @ * ! thesis and get a decent job."

All of these pressures combine to produce a continuous flood of journal papers, conference papers, technical reports, and news releases, many published months or even years before the reticent scientist would have liked. Once Pandora's box has been opened, once the Communities, not to mention the media and the politi-cians, learn of important or startling results, the scientist is very likely to lose control of the situation.

The Spectrum Revisited. The outputs of computer simulations arrive as neat tables of numbers, each with six or more digits, or as sets of neatly drawn graphs. Their format is the same whether the simulation model represents a mass/spring system or predictions of the mortality of presidents. The computer treats all input data the same way; it crunches numbers and generates neat output displays. It does not know or care whether the model was derived deductively or not and therefore has a high predictive validity or whether it represents a very black box.

In Chapter 3, we introduced the Spectrum of Models to illustrate the differing predictive validity of models and simula-tions originating in different fields within the physical, life, and social sciences. The validity of simulation models sharply limits or circumscribes the uses for which they are appropriate. In Figure 20, the spectrum is reproduced along with indications of the "legitimate" roles of the simulations.

In the very light end of the spectrum, simulations are impor-tant and invaluable tools for design, particularly for those of electrical and mechanical devices. In the design of signal pro-cessors, for example, simulations facilitate experimentation with

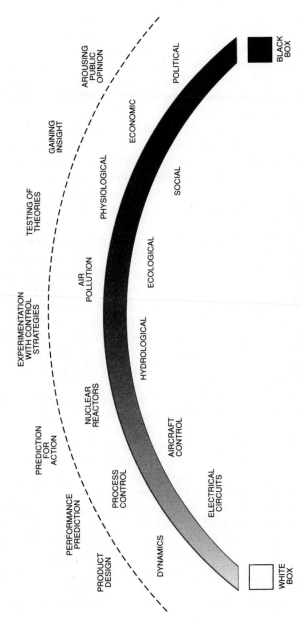

FIGURE 20. "Legitimate" uses of models and simulations.

various filter configurations and the selection of a circuit having optimal characteristics. Because the simulations are known to be very accurate, the designer can be confident that the outputs of the actual system, when built, will be exactly the same as the simulation outputs.

As the models take on a slightly gray hue, the simulation outputs are still sufficiently reliable to predict the performance of systems with considerable confidence. For example, pilots and astronauts are trained on simulators that are able to imitate the airplane or space vehicle so closely that the trainees are soon convinced that they are flying the real thing. The illusion is that faithful to reality.

Moving toward the middle of the spectrum, the models are no longer able to match reality output for output. For example, in simulating urban air pollution, the simulations cannot be counted on to predict smog levels on a day-to-day basis. There are too many inductive links in the causal chain. The simulations are very valuable, however, in predicting trends and in facilitating the evaluation and comparison of alternative mitigation and control strategies. For instance, local governments can explore the effect on average smog levels of curtailing or shutting down certain factories or traffic arteries.

This changes when we reach the darker regions of the spectrum, those that include models from biology and medicine. Here only rough or vague predictions are possible. Simulations now serve more for the study and exploration of the implication of various theories, rather than for the reliable prediction of events. In pharmacology, simulations are often used to study the probable effect of drugs proposed for the treatment of specific ailments and to help explain observed symptoms. In the past, laboratory animals were primarily used for such experiments to that end. With the mounting concern for the welfare of animals, simulations are providing a more humane alternative. It is very unlikely, however, that physicians would prescribe a medication solely on the basis of simulations.

Reaching the dark regions of the spectrum, the quantitative

significance of simulation outputs becomes more and more questionable. However, simulations can be very useful in providing qualitative insights into system behavior. Complex systems often contain interlocking feedback loops. These may cause the system to respond in a manner that is counterintuitive, the opposite of what one might expect. For example, in controlling the economy of cities, increasing the tax rate may actually result in a shrinking tax base, as businesses choose to relocate to the suburbs where the tax climate is more favorable. A well-designed simulation can help clarify this phenomenon. Similarly, strategy games and war games that include system simulations are often used to train corporate, government, and military decision makers.

In the very black extremity of the spectrum, the simulation results are almost purely speculative—stimulating subjects for conversation and for the imagination. They are sometimes used to shape public opinion, for advertising, and for propaganda.

So the way that we deal with simulation outputs must be carefully tuned to the location of the model along the spectrum. Unfortunately, the reports of simulations that reach not only the public but also the members of Communities and even experts rarely highlight the probable predictive validity of the models on which they are based. This has been the source of great confusion and misunderstanding in almost all application areas. Perhaps we need a law compelling all reports of simulations to bear a label such as "Warning! Use of these predictions may be damaging to your intellectual health and welfare!"

PREDICTING CATASTROPHES

Finally, we are ready to talk about the role of models and of simulations in predicting the catastrophic events—happenings so cataclysmic that they have a major, disastrous impact on society at large. We have carefully restricted our use of the term "catastrophe" to events which are both huge in magnitude and extremely rare. This makes their prediction by computer simulation especially subject to many of the pitfalls discussed earlier in this chapter. We need to say more about two of the most important sources of error.

Going Out-of-Bounds Revisited. Consider again our old friend the mass/spring system. We saw how a highly valid model could be formulated by starting with Newton's Laws and by using deduction. And we saw how inductive methods could be introduced to model nonlinear spring characteristics by experimenting with actual springs. So far so good. But we know that all springs wear out sooner or later. They are subject to fatigue and eventually fail by breaking into two pieces. So we must expect the mass to fall down at some unknown instant of time in the future. Let's call this event a minicatastrophe and use it to raise a very important issue. Can we predict at what time or after how many bobs up and down this minicatastrophe will occur?

We must recognize, first of all, that the model of the mass/spring system that we have discussed up to now is completely useless if we want to predict when the spring will fail. We are out of the bounds of validity for that model. Instead, we are confronted by a totally new and different phenomenon—metal fatigue. We cannot learn anything about this phenomenon inductively, by observing the mass/spring system while it is functioning normally. For this we need a totally different model.

If we want to find such a model, we must go to a specialist in metal fracture. Springs fail because the tiny metallic crystals of which they are constituted slide and slip. Perhaps we need a metallurgist or perhaps a structural engineer. Such a specialist will tell us that the susceptibility to failure of a spring is affected by microscopic nicks in its surface. The spring's precise chemical composition and the way it was hardened during manufacture also play a role. It may be possible, using X rays, to monitor the spring while it is being used and perhaps to give a brief advance warning when it approaches failure. That is the best the metallurgist can do for us, but that is not the kind of prediction we want; we may want to know long in advance.

If we have many springs available for study, we might set up an experiment to determine how many bobs each of, say, 1000 springs can take before it breaks. This may give us a statistical handle. We may come up with a probability; for example, we might be able to say that 99% of the springs survive 1000 bobs; 50% survive 10,000 bobs; and only 1% survive 100,000 bobs. Such a probabilistic model might be very useful for a manufacturer of bathroom scales or the like. But not for the kinds of catastrophes that we are talking about in this book. We cannot go out and run experiments. The catastrophes are too big and are not under our control.

All of the catastrophes that we will discuss in Part II of this book are huge. They are so large in magnitude that we can be sure that some of the variables of any model we propose will go out-of-bounds. The models that we have constructed and validated over the years are certain to become invalid with the approach of the catastrophes. And we have no way of validating models of out-of-bounds situations. In not one of the cases is there a law or a theory that we can count on. Nor is there the opportunity to run a series of destructive tests, as in studying the failure of springs. Nor can we gather statistics or use data from the past for this. The catastrophes are too rare. Most have never happened before.

When the variables become very large, approaching the bounds, "feedback" effects sometimes come into play. Two systems or subsystems, A and B, form a feedback loop, if the output of A is an input of B, and if an output of B is an input of A. Any change in the output of A is therefore fed back to the input of A. Feedback often reduces the rate of increase of system variables as the variables become larger; in some instances feedback may prevent the variables from ever reaching the point of catastrophe. For example, in the ecology of enclosed habitats, as the population of some animal species approaches catastrophic levels, the population of predators also increases and serves to place a cap on the population growth. But feedback may also make things worse rather than better.

In the mass/spring system, as the spring is compressed when

the mass moves upward, it reaches a limit beyond which it becomes extremely stiff, almost impossible to break. But when the motion is downward and the spring is stretched, it becomes weaker and weaker and eventually breaks. Hence, we have positive and negative feedback in one simple system. In large and complex systems, many feedback phenomena are operative and could act either to mitigate or to exacerbate the catastrophe in the "out-of-bounds" region. But, unless we have had a chance to observe these effects in action, we have no sure way of predicting how they will combine and affect the behavior of the system. They may help or they may hurt. We are very much in the dark.

Part II of this book is devoted to a detailed look at the eight predicted catastrophes that seemed to get the most attention and caused the greatest public concern at the time of this writing. A separate chapter is devoted to each of these imminent catastrophes. In six of the eight instances, the catastrophe is the direct result of a critical variable becoming steadily larger over time, until intolerable things happen. That variable is different for each of the six catastrophes—average global temperature, concentration of atmospheric pollutants, incidence of HIV infections, world population, and so forth. The variable has been too large for comfort for some time now, and in each case it is predicted to grow exponentially. Like a bank account earning compound interest, it is growing at an increasing rate. Each year's earnings are greater than they were in any preceding year. Except, of course, that we are not talking about a nest egg but about something disastrous. And in each of the six imminent catastrophes, the curves head off into the wild unknown. Eventually, the variables are predicted to stop increasing, to level off, but long after they have reached catastrophic proportions. This is illustrated in Figure 21.

Scientists have developed valid models that confirm what has been observed in the past and that are pretty good at predicting what will happen a little way into the future. Yet, when the variables are predicted to grow to magnitudes much greater than they have ever been before, there is then no way to validate the model, and the simulations go out-of-bounds in a big way.

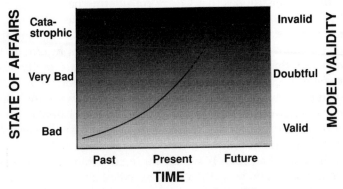

FIGURE 21. Typical catastrophe prediction.

The Curse of Dimensionality Revisited. Two of the catastrophes discussed in Part II, while very rare, do recur from time to time. In those instances, can we learn from the past? Can observations of past catastrophes be used to predict future catastrophes? In particular, earthquakes seem to recur periodically in some parts of the world. And economic crashes and depressions seem to exhibit a cyclical pattern. If we regard the death of a sitting president to be a catastrophe, we can use the presidential mortality example to illustrate the inherent problem with that kind of prediction.

We saw that from 1860 to 1960, all presidents elected in years that are multiples of 20 died in office—a remarkably consistent input/output relationship. On that basis we might regard presidential mortality to be a cyclical phenomenon and predict misfortune for the presidents elected in 1980, 2000, 2020, etc. But, as illustrated in Table 4, the year of election is only one of many attributes (parameters) that could be listed for each president. Lacking a plausible causal model, there is no way of telling which of these parameters, or combination of parameters, is a significant measure of the likelihood of death while in office. History really does not teach us anything in this case.

In the catastrophes that we are considering here, things are not as dark as all that. We have some ideas as to what might possibly cause an observed output, but for large systems in the darker regions of the spectrum, there are always hundreds of parameters that could be significant. Even if we are pretty sure that some specific parameter or input has an effect on the output, we can tell whether or not it is effective only if some of the other parameters assume certain values. That is, the correlation between an input and an output that we have observed in the past may hold only under certain, unknown conditions. For example, it may be that the year of election is significant in suggesting presidential mortality only if there is also a financial crisis and if the president is a heavy smoker. The system is complex, its dimensionality is too great to permit the exploration of *all* possible combinations of parameters—the curse of dimensionality. That curse actually haunts all of the models and simulations discussed in Part II.

In Conclusion. We have seen that models and simulations are subject to many sources of error, each of which may have a negative impact on the validity of the predictions based on the models. Some of them are under the control of the scientist doing the modeling; these usually involve trade-offs between additional detail in the model and the time needed to generate outputs on the computer. Other sources of error are beyond his control; they are inherent in the modeling methodology of each specific discipline.

The spectrum of mathematical models is useful in illustrating the relative validity of the models used in different application areas. It is very important to recall that the "shade of gray" assigned to different disciplines refers to the best models available in each discipline, those that carry deduction as far as possible and use reliable data when induction must be used. As the bounds or limits of any model are approached, the validity of the model necessarily suffers. The laws or principles that served as starting points for logical deductions are no longer guaranteed to hold; good inductive data become harder and harder to gather.

When we try to apply models to the prediction of catastrophic events, we are by definition out-of-bounds. Catastrophes always cause some of the system variables to become inordinately large, beyond the range of any previous measurements or observations. This means that, regardless of their validity (their "shade of gray"), in the normal operating range of systems all models become "black boxes" when pushed to catastrophic levels. This is just as true of mass/spring systems as it is of the presidential mortality model. Both models are essentially worthless when it comes to predicting the next catastrophe.

It may be argued that the definition introduced for the term "catastrophe" is overly narrow. After all, there are all kinds of unfortunate happenings ranging continuously from bad to worse to terrible to disastrous to catastrophic to "the end of the world." And simulations are very useful in dealing with most of them. Predictions may become gradually more unreliable, but they are still useful in that they may serve to alert us to the possibility of a catastrophe, even if they cannot predict just when it will happen.

Why should we focus so intently on the extreme situations for which the simulations are no good? The reason is that it is precisely the prediction of the extreme, the catastrophe, that gets the attention of decision makers and of the public. It is the prediction of a catastrophe that brings about a flurry of hectic activity. It is the prediction of a catastrophe that produces a reordering of priorities. There is a threshold beyond which the reaction of society takes on a qualitatively different form, not merely a quantitatively larger response. And the term "catastrophe" designates that threshold.

All disclaimers, restrictions, and qualifications aside, the bottom line is this:

Catastrophes cannot be predicted scientifically

PART II

Eight Imminent Catastrophes

Ashes to ashes, dust to dust—
If the lions don't get you,
The alligators must

—NURSERY RHYME

INTRODUCTION

Here we will look more closely at eight specific catastrophes that many scientists consider to be imminent. The warning of these catastrophes contributed to the information stress to which we were subjected in the early 1990s. They constitute a snapshot of what was considered potentially catastrophic at that point in time—and most continue to have catastrophe status to this day.

I have selected these eight catastrophes with some care. Each of the predictions or scenarios presented has the support, if not the total approval, of the scientific establishment. In each case there are very urgent recommendations for large government and private expenditures necessitating a major reordering of social priorities. Each one conforms to the paradigms and the methodology of a major scientific discipline, and each has been examined, debated, and evaluated by leading and respected scientists. We are, therefore, dealing with mainline thought. These are forecasts advanced by influential individuals who can count on respectful hearings from congressional committees, government agencies, and the media. We are definitely not dealing with fringe groups who spin pseudoscientific or far-out ideas.

It is the primary purpose of this part of the book to provide a perspective of what are perceived to be the major threats of our time, as well as the bases for the predictions of eight catastrophes in the near future. This is not an attempt to evaluate the various theories and data gathering efforts. We will accept these more or less at their face value. Nor are we interested in rank ordering them to determine which catastrophe is the most catastrophic.

In preparing this book, I consulted many of the original sources and references in each of the subject areas. This was interesting and educational. However, I have chosen to base each of the following eight chapters primarily on one or two books, books written in a popular vein and widely circulated in 1990. Each of these books was written by a respected scientific researcher or science writer. All received a measure of nationwide, even worldwide attention, and all have exerted an influence on the national and global decision-making process. I have kept away from hysterical, polemical, or self-serving publications.

With one exception, that of earthquakes, all of the catastrophes discussed below are global in character, threatening all or most of humanity. In all other respects, I have limited myself to the discussion of those predicted events that fall within the domain that I have defined as "catastrophic" in Chapter 1.

In summary, the attributes which characterize the catastrophes discussed in the following eight chapters are:

- Very unfavorable
- Extraordinarily intense in magnitude
- Exceedingly rare in occurrence
- Very long-lasting in its effects
- Imminent—expected in a few years or at least in a few decades
- Recognized by the scientific establishment as posing a very serious threat
- Brought to public attention by widely distributed books which were on the shelves of general bookstores in 1990

Within these constraints, I believe that I have not been guilty of serious omissions. In other words, the catastrophes discussed in the following chapters are not just examples, rather they constitute *the* major imminent catastrophes as generally perceived in the early 1990s. To set the tone, I have started each of the eight chapters with a brief quotation from one of the major recent books on the subject.

A Personal Note. Each of the eight chapters comprising this part of the book closes with a brief personal note. The major portion of each chapter is intended to constitute a succinct and relatively uncritical summary of the nature of the imminent catastrophe and of the judgments and recommendations of the pertinent Community. However, in the final section of each chapter, I have included my personal reaction to the inferences drawn from the models employed to predict the future course of events and to the mitigating actions that are urged by the Communities. My point of view clearly rests on subjective value judgments.

To some extent all expository writings reflect the bias and the parochial points of view of the author. In Part I of this book, I have tried hard to minimize the influence of my own educational and professional background and personality. Even so, I am painfully aware that the fact that I am a physical scientist, rather than a life or social scientist, has imparted a definite and unintended slant to the discussion. I have tried to be as objective as possible. By contrast, in the final section of each of the next eight chapters, called "A Personal Note," I make no such attempt. There I present my undiluted personal opinions. Therefore, it appears appropriate at this point to provide the reader with a brief overview, "a thumbnail sketch" of the way that I look at the world and how I view major social and political problems.

My early college education was in engineering and my first jobs involved geophysical prospecting for oil. I have been on the faculty of the University of California for some 35 years, most of that time in the Computer Science Department. My research has emphasized the use of computers to implement models of real-world systems, primarily in the physical science areas. I have consulted for a wide variety of industrial and government organizations and I have been very active in technical and professional societies.

Politically, I consider myself to be middle-of-the-road: on the liberal side of social issues and on the conservative side when it comes to economics. More of an interested observer than an activist. I tend to favor the environmental side in many controversies, but I try not to overlook or minimize the unfavorable economic consequences of proposed measures. I place a very high value on human life and human well-being all over the world. For this reason, I part company with the so-called ecowarriors when they put nature ahead of people.

FIVE

Ozone Layer Depletion
The Hole in the Sky*

> *The depletion of the ozone layer, the greenhouse effect, and the global environmental crisis are the most important issues that this country will have to face in the next decade and the next century.*
>
> —SENATOR ALBERT GORE *writing in Roan*[1]

Until relatively recently, when environmentalists were troubled by ozone, their concerns were entirely focused on its role as a major component of urban air pollution. Following theoretical work in the mid-1970s that suggested that certain human activities may have a harmful effect on the ozone layer which exists in the stratosphere, interest in that aspect of the ozone problem gradually increased. While ozone may be a noxious gas at ground level, in the upper reaches of our atmosphere it plays a vital role in shielding the earth from deadly ultraviolet radiation. The thinner this shield, the greater the expected incidence of skin cancers and

---

*Each chapter in this part of the book is headed by a brief quotation from a recent book on the subject.

other health problems. In 1986, it was discovered that the concentration of ozone above Antarctica had virtually dropped to zero, creating the so-called "hole in the sky." When chemicals called chlorofluorocarbons (CFCs) were identified as the primary culprit in this dramatic and very sudden disappearance of ozone, scientific and public concerns reached a feverish pitch.

The subject of ozone layer depletion differs in a number of respects from most of the other potential catastrophes discussed in this book. For one thing, differences among the various groups of scientists as to the chemical and physical causes for the phenomenon had been pretty well sorted out by the end of the 1980s. Influential figures in the scientific, industrial, and political worlds all, more or less, agree that CFC and related chemicals are bad for the ozone layer, and what is bad for the ozone layer is bad for humanity. There is still a great deal of disagreement over exactly how increased exposure to ultraviolet light would be harmful, but most people subscribe to the theory that further ozone depletion would lead to a catastrophe of worldwide proportions. This has led to concerted worldwide political actions that are unparalleled among the political responses to scientifically predicted disasters.

Among the vast number of technical papers, scientific reports, and books dealing with the various aspects of the depletion of the ozone layer, three books written in a popular vein are particularly informative and readable. They are Sharon L. Roan's *Ozone Crisis*,[1] John Gribbin's *The Hole in the Sky*,[2] and *Global Alert: The Ozone Pollution Crisis*[3] by Jack Fishman and Robert Kalish. The first two include concise discussions of the scientific background and of the scientific detective work that led to the identification of the chemical sources of ozone depletion. Roan's book provides a blow-by-blow account of the political battles that eventually resulted in the banning of CFCs.

Ozone is a relatively rare form of oxygen in that each molecule of ozone contains three atoms of oxygen, while the garden variety of oxygen has only two atoms per molecule. That extra atom gives ozone some very special properties. At or near ground level, ozone is a very noxious component of air pollution. It causes respiratory difficulties for people and does harm to plant life as well as to industrial products ranging all the way from rubber tires to ladies' stockings. However, things are very different in the upper atmosphere.

Starting at about 5 to 10 miles up, where the air is very, very thin, sunlight causes some of the oxygen in the atmosphere to break up and then to recombine to form ozone. This phenomenon occurs up to an altitude of about 30 miles, so that an atmospheric layer about 20 miles thick and rich in ozone is generated. This is the *ozone layer*, and the portion of our atmosphere in which it is located is known as the stratosphere. Stratospheric ozone absorbs a large portion of the ultraviolet rays, which are part of the incoming sunlight, thereby effectively shielding the earth from this very dangerous form of radiation. The thickness and density of the ozone layer fluctuates widely throughout each day and from season to season and is different over different parts of the earth. However, it is usually thick enough to protect living things—but that appears to be changing.

The development of spacecraft and, particularly, satellites has made all parts of the atmosphere accessible to frequent and accurate measurements. Throughout the late 1970s and the 1980s, satellite measurements suggested that the amount of ozone in the stratosphere was declining substantially all over the world. It was difficult to be quantitatively precise as to this decline or depletion. But it became very clear that something undesirable was happening. Then, in 1986, the "hole in the sky" was discovered, worries escalated, and scientists worked overtime to gather data and to perfect their theories.

THEORIES

Why is ozone disappearing from the stratosphere? Many scientists have made suggestions and proposed causal models. There were heated arguments well into the 1980s. By now, however, there is a general agreement on the major culprits and the principal mechanism for the depletion of the ozone layer.

Chlorofluorocarbons have been on the market for over fifty

years. They are widely used in refrigerators, under the trademark Freon, to enable spray cans and fire extinguishers to emit aerosol sprays, to make industrial solvents and plastics, and for a host of other purposes. CFCs are, in fact, a multibillion dollar industry. They have become so popular because they are exceptionally inert and therefore safe. They are not toxic, not flammable, nor do they react chemically with most other chemicals. But that is a large part of the problem.

Because they are very light, CFCs when released (or sprayed) into the air, gradually rise up, up, up. Having nowhere else to go and not subject to chemical decay or removal, they accumulate in the stratosphere. During the forty-year period from 1940 to 1980, the amount of CFCs in that layer of the atmosphere had risen from negligibly small quantities to well over 300 tons—that is a lot of gas, and the amount keeps increasing.

In 1974, researchers Sherry Rowland and Mario Molina theorized that the ultraviolet radiation entering the stratosphere causes CFC molecules to break up and form free chlorine. This chlorine acts as a catalyst for a series of reactions that have the effect of converting ozone into oxygen. Since the chlorine in the CFCs is not used up in this process, the CFCs will keep eating up ozone as long as there is sunlight. The more CFCs in the stratosphere, the greater is the depletion of ozone. In addition to the CFCs, a number of other chemicals can cause ozone to disappear. These include industrial chemicals such as methyl chloride and methyl chloroform as well as natural sources of chlorine including the burning of plants and the eruption of volcanos. But CFCs are by far the most important cause. Tons of data have been gathered and published to confirm this.

But why is there a hole over Antarctica?

Actually the hole is seasonal. For the past several years, the ozone level in the stratosphere has dropped almost to the vanishing point every October, at the end of the Antarctic winter season, and has more or less recovered fully during the summer months. This phenomenon came as a complete surprise when it was first observed in the early 1980s, and its cause became a subject of heated controversy among scientists. But it is now more or less generally agreed that CFCs and other chlorine-containing chemicals are responsible for the hole. Because these compounds accumulate in the stratosphere during the long polar winter night, the very low temperatures near the poles make these compounds particularly damaging to ozone when ultraviolet rays appear there in early spring.

Now what about the effects of the depletion of the ozone layer on us? Everybody agrees that too much ultraviolet is bad for you and all living things. Were it not for the shielding provided by the atmospheric ozone, most life forms might well vanish from the land surfaces of the earth. But how much shielding is really needed, and how would people be affected if increasingly high amounts of ultraviolet gets through? Nobody is sure.

It is known that excessive doses of ultraviolet rays cause sunburns and skin cancers. It has been estimated that each 1% reduction in the amount of stratospheric ozone would increase malignant skin cancers by 1 to 2%. But there may be other significant effects on health including increased levels of herpes, cataracts, and the suppression of the activity of the human immune system. There is also a major impact on plant life, possibly putting many of our essential sources of food in jeopardy. There are also a number of theories tying ozone depletion to major changes in the climate, similar to those anticipated due to the greenhouse effect.

DATA

Even with scientific satellites and the latest instruments, scientists have extreme difficulties in coming up with conclusive data regarding the worldwide depletion of the ozone layer. The amount of ozone in the stratosphere varies greatly from place to place and fluctuates substantially from day to day, from season to season, and from year to year. Most of these fluctuations are due to weather and climatic patterns, solar activity, and other causes that have nothing to do with CFCs. But we now know that over the past two decades or so there has been a steady decrease in the average amount of ozone in the stratosphere—and the trend is continuing.

The hole in the sky can be explored more easily because here we are talking about a very specific geographic location, a very specific time of the year, and a relatively enormous change in ozone concentration. A depletion of the Antarctic ozone layer in excess of 30% of its normal range was confirmed by ground measurements and by weather satellites as early as 1984. By October 1987 that depletion level had reached 50%. Remember, however, that most of the ozone is restored during the Antarctic summer months. Other very accurate measurements established a strong correlation between the level of CFCs in the stratosphere and the decline of ozone concentration.

In 1986, the United States sponsored a National Ozone Experiment (NOZE), and in 1987 NASA led an international Airborne Antarctic Ozone Experiment. These experiments included measurements using instruments carried by balloons as well as by specially designed and equipped aircraft. Most of these very detailed and very costly measurements provided general confirmation of the theories linking ozone depletion with the continued and increased use of CFCs and other industrial chemicals.

THE CAUSAL CHAIN

So how do scientists put all these theories and data together to predict what will happen in the future? They have to use a model of a causal chain that starts with the CFCs emitted into the atmosphere and ends up with possibly catastrophic diseases and with the effects of changes in the world climate. Such a model is actually a chain or network of smaller models, each characterizing a part of the puzzle. There is first the model that describes the quantity of the CFCs that is emitted into the atmosphere, how these emissions are distributed over the earth, and how much winds up in the upper atmospheres where it can do its mischief. We designate this model as Model A in Figure 22, which pictures the entire causal chain in a coarse fashion. The output of Model A serves as the input of Model B, which describes the chemical processes in the upper atmosphere, the processes that cause the depletion of ozone.

Model B must furnish, as its output, not only the gross amount of ozone that disappears but also the geographic distribution of the ozone that remains—the location and the size of the holes in the sky. We know that the depletion of the ozone layer causes an increase in the amount of ultraviolet light that gets through to earth. Now we need models to describe precisely what harm this increase in ultraviolet radiation will do to people directly

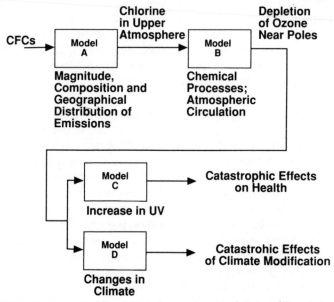

FIGURE 22. Comprehensive model of ozone layer depletion and its catastrophic effects.

and, on the other hand, indirectly by causing major changes in the world climate, changes that in turn produce catastrophic storms, tidal waves, and so forth. These effects are represented by Model C and Model D in the figure.

PREDICTIONS

Given the difficulties in determining worldwide average ozone levels, it is not surprising that predictions of the depletion of the ozone layer during the coming decades have varied widely. The prestigious and authoritative National Academy of Sciences has published predictions ranging from as low as a 2% depletion of the ozone layer to as high as 20% in the long run. Others have predicted depletions as high as 40%. By 1990, the accepted figures for the amount of ozone layer depletion since 1967 were 10% each for Europe and North America and 3% at the equator.

These are small, however, compared to the observed and predicted depletions (as high as 100% during October) over Antarctica. But why worry so much about the penguins and the few intrepid explorers that can be expected to suffer from that phenomenon?

For one thing, the hole in the sky seems to be growing. Recent measurements indicate a precipitous seasonal ozone decline all the way north to Australia and New Zealand. And predictions indicate that eventually the entire Southern Hemisphere will be influenced by the hole. It is predicted that this phenomenon will produce ozone depletions of 3 to 4% over and above what would occur if the hole were not there. Now there is evidence mounting of a northern hole, above the North Pole. If such a hole develops, it will have a very important negative impact on ozone levels in the Northern Hemisphere. To top it off, there have even been suggestions that the growing polar holes could set off a chain of events that could affect the worldwide climate and lead to the early and rapid onset of the next ice age, while we are worrying about global warming.

WHAT MUST BE DONE

Here there is very little disagreement. We must stop releasing chemicals that destroy the ozone layer immediately. This means, first and foremost, an end to all CFCs that are now used in most refrigerators, spray cans, and many industrial processes. It also involves the elimination or at least the decrease of the utilization of a host of other products that are composed of organic molecules containing chlorine. Bear in mind many of these products play a vital role in improving the standard of living, particularly in the Third World. So it's not just a matter of banning the use of this or that; acceptable substitute chemical products must be developed and introduced. Since this is sure to incur great costs, we must find the money to pay for it.

Even then, will the problem be solved? Not entirely. The chlorine that is already in the stratosphere will act as catalyst for a long time. Even if we stop releasing harmful chemicals tomorrow, there are many natural sources of chlorine, for example, volcanic eruptions. Still, it will be a significant step in the right direction.

POLITICS

Among the many potential catastrophes described in this book, the ozone layer depletion problem is unique in having elicited a relatively prompt and concerted response from governments all over the world. Recall that the possibility of a catastrophic depletion of the ozone layer was first seriously proposed by scientists in the mid-1970s and that it took years to establish conclusively that the ozone layer was indeed shrinking and more years to become convinced of the key role of the CFCs. It may seem surprising, therefore, that in 1987 representatives from fifty-six nations acted relatively swiftly by meeting in Montreal and agreeing to decrease the production and use of most CFCs by half by the year 2000. At the time it was considered to be a colossal achievement in international cooperation, but it was clearly only a promising first step.

When new and very disturbing data became available after the Montreal conference, representatives of an even larger number of nations around the world (including developing countries such as India and China) met again in London in 1990. There they signed a protocol to the effect that the five most important forms of CFC be phased out completely—reduced to zero by the year 2000. They added a number of other ozone-destroying chemicals to the list of products to be eliminated. To help the poorer nations of the world meet the expenses that this action entails, a fund in excess of one quarter billion dollars was set up, with more money to be allocated later. This amount is, of course, only a drop in the bucket compared to the billions of dollars that will have to be spent in the industrialized countries to develop CFC substitutes and a host of new chemical products. Truly this was "A Giant Step for Mankind," as more than one editorial writer put it.

A PERSONAL VIEW

In looking at a chain of models, such as Figure 22, it is important to recognize that this kind of cascading leads to a compounding of the errors inherent in each link of the chain. The probabilities must be multiplied. For example, suppose we think that the validity of Model A is 60% (or 0.6), meaning, roughly, that the model will predict the correct concentration of chlorine 60% of the time and that the validity of Model B is 10% (or 0.1). The validity of the output of Model B would then be 0.6 times 0.1, which is 0.06 or 6%. So the validity of the chain consisting of Model A and Model B would be only 6%. This is the validity of the inputs to Models C and D, the validity of the outputs of the latter two models would be only 6% of their individual validities. The point is that the greater the number of models connected in a chain, the worse the validity of the overall predictions.

As far as Model A is concerned, scientists feel that they have the problem of predicting the amount of chlorine in the upper atmosphere rather well in hand. The amount of CFCs that are released into the atmosphere each year can be estimated reasonably well. Moreover, chemists have studied the pertinent chemical reactions since the mid-1970s and many of their predictions have been confirmed by data gathered by satellites.

By contrast, the prediction of just how much ozone will be removed from the upper atmosphere by the chlorine and over what parts of the earth this depletion will be most pronounced is still pretty much an open book. Only recently has it been discovered that a hole in the sky has formed over the Arctic region, not just over the Antarctic. Clearly, winds in the upper atmosphere probably due to seasonal temperature variations have something to do with it. But theories to explain all this are still being devised.

The effects of ozone depletion and of the concomitant increase of ultraviolet radiation that gets through to the earth is even more of a mystery. This is true of the public health effects and climate modification. For example, it is known that exposure to ultraviolet radiation can result in skin cancer. But there are different kinds of skin cancer, some lethal and others not so dangerous, and the susceptibility to skin cancer varies widely from person to person. Hence, there is wide disagreement regarding the quantitative relationship between seasonal increases in ultraviolet radiation and the incidence of skin cancer directly attributable to these increases. There is even more uncertainty concerning the manner in which ozone depletion may affect the global climate. Some scientists predict increased global warming, while others believe that the effect will be to cool the atmosphere. There are theories, conflicting theories, and educated guesses; but there is no one generally accepted model.

We have a fairly valid model, linked with two models of questionable validity. The validity of the chain of the three models (Model A, Model B, and either Model C or Model D) is therefore very low—much lower than the worst one of the three. In other words, quantitative predictions of the effects of ozone layer depletion should be taken with a grain of salt (actually, a lot of salt). I would not venture to assign actual probability figures to any of the models (the numbers 60% and 10% cited above are purely illustrative). In Chapter 4, I pointed out some of the pitfalls inherent in the use of probabilities. On the other hand, we can be pretty sure of the general trends, the qualitative aspects of the predictions.

I am persuaded that the ozone layer is being depleted by chemical products emitted into the atmosphere. I agree that the resulting increase in ultraviolet radiation at ground level is harmful not only to humans but also to many animal and plant species. It also seems probable that the depletion of the ozone layer has undesirable effects on the climate. But how "catastrophic" are these effects?

The predicted long-term effects of increased ultraviolet radiation on the multitude of earth's life-forms are varied. Some species will thrive, others will adapt, but many will suffer. In 1991, the Environmental Protection Agency estimated that in the course of the next fifty years, the United States would see 200,000 additional deaths from skin cancer due to the depletion of the ozone layer. That's a lot of fatalities. But even if these estimates turn out to be low, the number estimated is but a tiny percentage of the deaths attributable to other diseases and other life-threatening hazards. Therefore, if the objective is primarily to save human lives, perhaps the money would better be invested in various medical programs.

The major application of CFCs is as coolants in refrigerators and this is also a great problem. Over 40% (by volume) of the CFCs used in the United States serve this purpose. In the developing countries, refrigerators are just beginning to be introduced in most households and are considered a significant component of improved standards of living. A decrease in the number of refrigerators in the developed countries and a slowdown in the introduction of refrigerators elsewhere is likely to result in severe increases of health problems and fatalities due to food spoilage, food poisoning, and even famines.

For this reason, there is a frantic search for suitable substitutes for CFCs. Present alternatives such as helium would require three or more times as much energy and are therefore unacceptable from the conservationist point of view. There is, of course, the hope that a new "wonder chemical" to replace the CFCs will be invented. But can we be sure that such a chemical will not eventually turn out to be harmful in some unsuspected way? After all, CFCs were widely used for almost fifty years before their effect on the ozone layer was suspected.

I fully support efforts to reduce and eliminate the use of CFCs and other ozone-depleting culprits. The threat is definitely here, and concerted international action can appreciably help to blunt its impact. But I am a bit troubled by Senator Gore's statement that the depletion of the ozone layer is one of the three most important issues that this country will have to face in the next decade and the next century. On my own scale of things, I certainly rank overpopulation (and the attendant problems of famines, decimation of forests, epidemics, etc.) and the AIDS epidemic, to which I have devoted separate chapters, ahead of ozone-layer depletion.

SIX

Climate Modification
The Greenhouse Effect

*We should be frightened for the survival of
life on Earth. We have a problem that is
serious, immediate, growing quickly, and is
potentially devastating. We need to wake up!*
—ALBERT K. BATES[1]

The term "greenhouse effect" is a picturesque, though perhaps
misleading, metaphor for global climate changes brought about
by increasing amounts of carbon dioxide (CO_2) in the atmosphere.
It is misleading because we ordinarily think of a greenhouse as
providing a climate favorable to the growth and well-being of
living things. But here there is concern that unchecked increases in
the average temperature all over the world could produce catastrophic effects.

Throughout the 1960s, and well into the 1970s, climatologists
believed that the earth was gradually cooling, and they recommended a variety of actions to help us adapt to the expected frigid
weather. Then it was discovered that the level of CO_2 in the
atmosphere had been rising steadily at an alarming rate for some
years. Theoretical conjectures as well as a large number of com-

143

puter models suggested that this increasing amount of CO_2 and other so-called greenhouse gases would raise the temperature disastrously. Recent measurements of temperatures all over the world have confirmed that theory, at least in the eyes of a majority of knowledgeable scientists.

There has been a spate of recent books on the greenhouse effect and related environmental problems. The most authoritative and readable of these is Stephen H. Schneider's *Global Warming*,[2] which also provides an insightful and balanced discussion of what climate models can and cannot do. In a more popular vein, *Hothouse Earth: The Greenhouse Effect and Gaia*,[3] by John Gribbin, is a very skillful and broad review of the subject.

THEORIES

Why should the world be getting warmer at this juncture in history? Is it really getting warmer? Is it the result of human activities? How can we tell?

Explanation for Warming. Greenhouse effect theories start from the observation that the amount of CO_2 in the air has been increasing rather steadily over the last thirty or more years. On a mountain top in Hawaii, the concentration of this gas has been carefully measured and recorded for over thirty years, since well before global warming became an issue. In the course of the thirty-year period from 1958 to 1988, the concentration of CO_2 increased from 315 to 350 parts per million or about 11%. These data started people thinking about the role of atmospheric CO_2 in determining our climate.

The Basic Idea. Earth is heated by sunlight. Some of the incoming energy is reflected by the upper atmosphere, some of it is absorbed by the air on the way down, but almost half of it reaches the surface of the earth and heats it up. The warm earth in turn radiates heat in the form of infrared waves. This energy in the infrared band is partially absorbed by the atmosphere and reradiated back to earth. It is this exchange of infrared that makes the earth considerably warmer than it would be were there no atmosphere. It is also this heat exchange that constitutes the greenhouse effect. In fact, without an atmosphere to catch and return the infrared waves, the earth would be about 35°C colder than it is now and would be a very inhospitable place indeed.

The amount of greenhouse warming and the average temperature of the earth's surface are determined in part by the composition of the atmosphere—the percentage of moisture and of various chemicals. CO_2 is so important in this composition that scientists have been able to demonstrate a direct relationship between the temperature and the level of CO_2. In particular, a good deal of work has been performed to analyze conditions during the ice ages that have occurred on our planet periodically.

Geological records show that over the past million years, the earth has been quite cold most of the time, except for so-called interglacial, warm periods that last about 10,000 years and recur every 100,000 years, more or less. Analyses of core samples obtained by drilling very deep holes into the polar ice caps show that the interglacial periods were accompanied by relatively high levels of CO_2 in the air. Hence, CO_2 and elevated earth temperatures seem to have gone hand in hand for a long, long time. Actually, CO_2 is not the only gas that absorbs infrared radiation. Other components of the atmosphere, such as CFCs, methane, nitrous oxide, and ozone (at low altitudes) are believed to contribute to the greenhouse effect.

Where do all these greenhouse gases come from? Most are the result of natural processes, processes that were in play long before man came along. The amount of greenhouse gases in the air has fluctuated in daily, seasonal, and long-term cycles for at least a billion years. But the amount of CO_2 that has recently been measured, and the increase in its concentration during the past thirty years is unprecedented. Most scientists agree that this can only be ascribed to recent human activities, including the increased use of fossil fuels, such as coal and oil, and to the worldwide decimation of forests by logging and the clearing of tropical rain forests for agriculture.

Climate Models. Climatologists have been hard at work for decades to develop mathematical models of the climate and to implement these models on the most powerful available computers. The models that they have come up with vary widely in their complexity and in the variables that they are designed to predict. The most advanced and sophisticated of these models are called *general circulation models* (GCMs). Here is a very rough idea of how they work.

Think of the entire surface of the earth covered by enormous cardboard boxes laid side by side, so every point on land or sea is under a box. Each box measures several hundred kilometers in the north–south and in the east–west directions, and each box is about three kilometers high. On the top of each box is another identical box, forming a second layer of boxes, and to this are added more layers. Hence, there are layers upon layers of boxes, actually about ten layers in all, extending to about 30 kilometers above earth's surface—hundreds of thousands of boxes. A typical box has six adjacent boxes (four in the horizontal and two in the vertical directions).

Before starting the simulation, the scientist specifies and reads into the computer a set of variables for each box—e.g., temperature, air pressure, moisture content, wind velocity and direction, the concentration of all chemicals, and many others. These are called the *initial conditions*. In addition, the scientist specifies and programs into the computer a set of transient expressions describing all the things that he expects to flow into and out of the atmosphere after the start of the simulation—variables including the solar radiation, evaporation from the sea, rainfall and snow, heat radiated by the earth, chemicals such as air pollutants, and many others. These are the system inputs and are called the *boundary conditions*.

Already in the computer is a very elaborate set of equations that describes how matter and energy move from box to box as time goes on. When the actual computations begin, the computer uses all this information to calculate the magnitudes that all variables in all the boxes will assume, say, six hours after the time for which the initial conditions were originally specified. These new values for all the variables become a new set of initial conditions for the next step in the computation. This process is continued until, say thirty or fifty years of actual time have passed. Obviously, this entails an immense number of arithmetic operations. Such a computer run can easily take several hundred hours on the fastest and most expensive of our supercomputers. These runs must be repeated again and again for different combinations of initial and boundary conditions. Finally some output variables, such as the temperature at selected locations, are displayed.

As described in Chapter 4, all simulations of this type are subject to a great many serious sources of error, some fairly obvious and others more subtle. First, there are the uncertainties about the equations that govern the flow of matter and energy from box to box. Then there are the uncertainties about the boundary and initial conditions. Also, even with hundreds of thousands of boxes, the simulation of the atmosphere is never sufficiently detailed. Environmental features such as clouds, for example, cannot be represented adequately. In addition, there is the problem of error accumulation as the solution is stepped through time. Finally, there is the chaos problem. One hopes as time goes on, as more and more is learned about the system and about the models, that the computer outputs will gradually become more meaningful and credible. But it is always extremely difficult to tell whether one is looking at significant system outputs or a potpourri of errors.

The Consequences. If it is hard to forecast future temperature trends, it is even harder to predict the effects that such trends would have on the environment. One of the most damaging effects of persistent global warming could be a catastrophic rise in the sea level. It is not necessary to postulate a melting of some or all of the polar ice caps or of glaciers. As the average atmospheric temperature rises, ocean waters expand—and that makes the water level go up. In the past, it appears that a warming by 0.5 °C eventually produced a rise in sea level of about 10 cm. But all kinds of uncertainties still abound.

For instance, the temperature of seawater is nonuniform over the earth. Also, air temperatures have a direct effect on water temperatures, primarily on the water near the surface, not at great depths. But it is clear that the seas will rise, which in turn will threaten land areas, cities, and installations that are located close to sea level now. How much of a rise is open to conjecture.

There are also other indirect consequences to be expected from global warming. It is theorized that instabilities, i.e., changes, in the average temperature make a variety of climatic extremes more probable: scorchingly hot summers, extraordinarily frigid winters (even as the average global temperature increases), intense storms, droughts, floods, earthquakes, volcanic eruptions—the list goes on and on. It all depends on which model and which scenario you use. From there you can conjecture on the resulting effects on agriculture, economics, health, and society in general.

DATA

As already intimated, it is extremely difficult to know which measurements will permit the calculation of average "global temperatures"; even the precise meaning of the term cannot be agreed upon. We are looking for average changes of a very few degrees over periods of one hundred years; this while seasonal and even daily changes in temperature are often more than twenty to fifty times greater than the average temperature changes. And, of course, we can only record the temperatures at a limited number of measuring stations. Furthermore, what are we to make of scattered reports of temperatures measured fifty or even a hundred years ago with the primitive instruments then available?

For sure, you cannot recognize trends in the climate by looking out of the window. Our judgments must not be overly influenced by one hot summer or series of hot summers, nor by exceptionally cold winters. These are deceptive, not only because they may mask or appear to overshadow long-term trends in the opposite direction, but also because both hot and cold extremes in temperature are to be expected simultaneously when the climate is changing—whether there is global warming or global cooling. Yet, unusually hot summers or cold winters seem to get a disproportionate amount of attention by the media.

One of the most widely quoted and admired sources of global temperature trends is NASA's Goddard Institute for Space Studies (GISS), where James Hansen is in charge of a team of scientists that gathers many kinds of data regarding the climate and corrects or adjusts these to take into account many different sources of error. The team has plotted the temperature deviation over a one-hundred-year period. The temperature fluctuations are fairly small, of the order of a few tenths of a degree. But the trend appears to be rising. Other research units, using different data and different data reduction methods, have come up with similar results.

With ocean currents, storms, dams, evaporation, and the like, sea level changes are even more difficult to calculate. When the global sea level is plotted against time over the past hundred years, a rising trend emerges that closely follows the rising temperature. This lends support to the theory that the temperature-induced rises in sea level are due to the expansion of water (which takes place quickly) rather than to the slow melting of ice caps, since the effect of the latter would tend to lag behind the temperature rise by several years.

There is less uncertainty about the amount of CO_2 and other greenhouse gases in the atmosphere, since they can be measured quite reliably by modern instruments. The ice core samples, already mentioned, demonstrate a close correlation between CO_2 levels and temperature over the past 160,000 years. Furthermore, it is quite clear that present-day CO_2 levels are higher than any that have existed over that entire span of time. There simply is no explanation for this extraordinary buildup of greenhouse gases other than the greatly increased human activities in the second half of this century. Everybody seems to agree on this.

THE CAUSAL CHAIN

There are many greenhouse gases and a myriad of sources of these atmospheric pollutants. Some are caused by natural processes such as the transformation and decay of animal and plant matter; others are the direct result of human activities such as emissions from power plants, automobiles, etc.; still others are indirectly caused by people, such as the emission of methane from cattle that are raised for human consumption. In Figure 23, the varied sources are encompassed by Model A. The outputs of Model A feed into Model B, which determines the rise in the global temperature that is produced by the greenhouse gases. The increase in global temperature is expected to produce major changes in the earth's climate. These are mapped by Model C to predict the specific unpleasant consequences of the climate modifications: rise in sea level, droughts, and storms, among others. Models D, E, and F translate these into possibly catastrophic effects on society at large.

It may be worthwhile to repeat once more a generalized summary of the discussion of chaining in Chapter 4. If each of the models in Figure 23 has a validity that may be estimated, the overall validity of the comprehensive model is found by multiplying the validities of the component models forming a chain. If, for

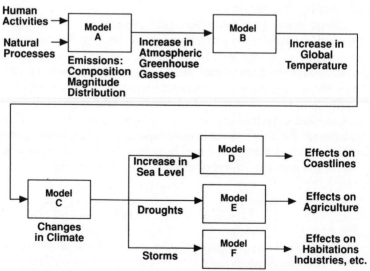

FIGURE 23. Comprehensive model of the greenhouse effect.

example, each of the Models A, B, C, and D were assumed to have a validity of 0.2 or 20%, the validity of the predicted effects of rising sea levels would be $0.2 \times 0.2 \times 0.2 \times 0.2$, which is 0.0016 or about one sixth of one percent.

Were we interested in coming up with a quantitative characterization of the validity of the comprehensive model shown in Figure 23, we would have to estimate or guesstimate the validities of each of the component models. That would not be a very meaningful exercise. However, we can look at some of the models a bit more closely, realizing that each of them is actually a coarse representation for a vast network of smaller models, of submodels.

Take Model A, for example. We can make a pretty good estimate of the amount of carbon dioxide that is currently injected into the atmosphere by factories, power plants, cars, etc. We can also take a stab at estimating the CO_2 that is emitted into the atmosphere and removed from the atmosphere by plants, especially by trees. But a lot more carbon dioxide and methane is generated by the natural processes that cause dead plant and animal matter to decompose. This is very hard to estimate and may, in fact, depend on the temperature, the amount of global warming. Hence, referring to the spectrum of models of Chapter 3, Model A by itself is fairly dark gray.

Model B is comprised of a host of physical and chemical processes in the atmosphere that we are just beginning to understand. There are many controversial features. Referring to the general climate model (GCM), one of the best available to us, Stephen Schneider[4] states that: ". . .validation examples provide strong circumstantial evidence that the current modeling of the sensitivity of global surface temperature change to increases in greenhouse gases at least over the last 100 years or so is probably valid within a factor of 2 to 3. The validation of regional predictions is more problematic. . . ." That is how accurate the models are in "predicting" how average global temperatures have changed during the past century. The future will surely take us into temperature and CO_2 ranges with which we have no experience, which are "out-of-bounds." So Model B is very, very dark gray. And Models C, D, E, and F are even more problematic. This suggests that the validity of the comprehensive model is minuscule at best.

PREDICTIONS

Most forecasts of global warming and the accompanying side effects are followed by sets of prescriptions of things that should be done to avoid the most catastrophic consequences. The predictions, therefore, tend to deal with what may happen if we fail to mend our ways and with what is to be expected if we take some preventative actions—many best-, medium-, and worst-case scenarios. These scenarios require at least three distinct steps.

Having fairly well established that certain greenhouse gases, particularly CO_2, are responsible for the greenhouse effect, we must predict how the levels of these atmospheric gases will change in coming years. Then we must predict how this will affect worldwide, and possibly local, temperatures. Finally, it is necessary to predict the nature and magnitude of the unpleasant consequences of these temperature changes.

There is tremendous uncertainty as to the amount of the buildup of CO_2, let alone all the other greenhouse gases, to be expected in the coming decades. It depends strongly on the predicted growth in the developed as well as the developing world, the extent to which energy can be economized, and the extent to which fossil fuels can be replaced by alternative sources of energy. Certainly, how we treat or mistreat our forests will also have a significant effect. Other sources of greenhouse gases include "natural" phenomena such as the decomposition of dead organic matter and the release of CO_2 from carbonate sediments in the oceans. Hence, all reputable prognosticators hedge their predictions. However, there is general agreement that unless we take very drastic actions at once, sometime in the first half of the twenty-first century the level of atmospheric greenhouse gases will be twice what it is now. The only question is how soon.

In order to predict the effect on the temperature of the buildup of greenhouse gases, one has to specify or guess the significance of various negative and positive feedback processes and make a host of other assumptions. These are too complex to include their discussion at this point. The bottom line, however, is that most authoritative scientists venturing predictions place temperature increases between now and the year 2050 at about 0.3°C (or about 0.2°F) every ten years, if there is no concerted effort to reduce the emission of greenhouse gases. Others put this figure above 0.5°C per decade. If strong mitigating actions are taken, it may be possible to hold this increase to 0.1°C per decade. This would mean an increase in the world average temperature anywhere from 0.6°C to 3.0°C. These are global averages. The prediction of regional or local temperature changes is much tougher still.

Stephen Schneider[5] lists the following consequences of global warming over the next several decades:

- Increased rainfall in the tropics
- Wetter springtimes in high and middle latitudes
- Drier midsummer conditions (droughts) in the United States
- Increased probability of extreme heat waves
- Increased likelihood of forest and brush fires
- Increase in sea level by as much as a meter over the next hundred years

Others would add a sharply increasing number of blizzards, hurricanes, dust-bowl conditions (in the United States), and many other disasters.

WHAT SHOULD BE DONE

Every environmentalist has his own list of remedies, but by and large they all contain the following major items:

- Sharply increased funding for research; systematic large-scale programs to develop a better understanding of climatology, as well as the many problems related to the reduction of emissions of greenhouse gases.

- Reduced energy consumption by all segments of society.
- More efficient use of energy.
- Shifting from coal to less harmful fossil fuels, such as natural gas.
- Phasing out harmful chemicals—first CFCs and gradually fossil fuels, at least to some extent, by developing and introducing alternative sources of energy.
- Planting a large number of trees and preventing the decimation of the tropical rain forests.

All this will cost lots of money. Schneider[6] proposes an annual expenditure equal to "5% of the defense budget each year to the year 2000 A.D." to finance the development and implementation of installations in line with the above list.

While environmentalists strongly favor the curtailing of fossil-fuel power generators, they are also nearly unanimous in condemning one obvious source of energy that does not emit greenhouse gases—nuclear power. No doubt, fear of pollution as a result of nuclear reactor accidents and of improper radioactive waste disposal is the basis for their opposition to fission reactors. High development costs are usually cited as the principal objection to yet-to-be-invented reactors that operate by nuclear fusion.

A PERSONAL VIEW

As I see it, the detailed predictions of global warming are based on models that contain many influential ingredients that are only partially understood, at best. The effect of cloud cover on global warming is still a controversial issue, and the yet-to-be-determined role of the oceans in regulating the temperature may be very important. Moreover, the average global temperature is apparently exceeding all past temperature maximums. The climate models are, therefore, being exercised beyond the bounds of their established validity. We are in unexplored territory as far as these models are concerned, and we can be sure that the chaos phenomenon is also going to be a factor. Hence, in terms of the discussion in Chapter 3, global climate models are in the very dark region of the spectrum. Their credibility as predictors of specific events is very, very low.

Let me hasten to add, however, that it would be wrong to conclude that the models and submodels of the greenhouse effect are no good, that they are useless. In fact, most decision making by the government and by all of us is of necessity based on models that are far less elegant, plausible, or meaningful than are the climate-modification models. Life is rough and so is system modeling. We must do the best we can with the modeling concepts and tools that are available to us. They are our best hope for learning to understand and perhaps to control our environment. But we must take great care not to be overly impressed and misled by slick-looking computer printouts and computer graphics that suggest that the quantitative predictions have a greater significance than they really do.

What are we to make of specific forecasts of future global temperatures? We should not regard the predictions as prognoses of what is likely to happen or what will actually happen. In the year 2050, we should not plan to compare the predictions made by each model to find out which one of them has turned out to have been the best. There are so many potential sources of error in each model that coming close to the correct temperature is most likely due to a fortuitous compensation of errors, many mistakes canceling each other out. However, global climate simulations greatly improve scientists' understanding of the phenomena being simulated. Even if the simulations are not quite true to nature, they greatly enhance our ability to make good choices from alternative strategies for controlling the processes being simulated. As the noted mathematician Richard Hamming observed: "The purpose of simulation is insight, not numbers." Insight is the key. Because of the very extensive large-scale simulation efforts that have been under way for well over ten years, we now know much more about global climate change than we did before.

Hence, while we can quarrel about numbers, the general trends are clear: greenhouse gases, especially carbon dioxide, have been accumulating in the atmosphere at unprecedented rates. Human activities are largely, though certainly not wholly, responsible. These greenhouse gases are causing the globe to become warmer and warmer. We are definitely in an era of global warming and we must expect the average global temperature to rise appreciably over the coming decades. Now, how catastrophic is this?

Very few scientists assert that the world would be a less hospitable place for humans and most other life forms if the average temperature were to be a few degrees higher than it is now. Some places that are pleasant now would become too hot or too dry. On the other hand, some regions that are now too frigid for comfort would become more habitable. Some areas that are now close to the beach would be flooded; but somewhere else useful land surfaces would be exposed as ice caps and glaciers recede. In the long run, most human activities and most living species could adapt.

The problem with global change is in the short run—the transients, the things that happen while the temperature is changing. If arid conditions due to global warming make the American midwest less suitable for agriculture, it may take many decades before newly fertile regions in the Canadian north and in Siberia can take over as breadbaskets for the world. In the meantime, famines would be a real possibility. Droughts in other parts of the world may force people to turn to polluted water supplies, and this may result in various epidemics. Rising sea levels would also pose severe problems. For instance, people living along the shores of Bangladesh could not rapidly be transplanted to Canada or Russia. New cities would have to be built. New transportation systems and health delivery systems would have to be created. In fact, the entire world infrastructure would need adjustment and such adjustments take a long time. In the meantime, there may be immense suffering.

Therefore, the problem is really not change per se, the problem is rapid change. Change so abrupt that usual mechanisms for adjusting to change cannot keep up. The climate has always been changing. In the postscript to this chapter we will learn that the world is unquestionably heading toward an ice age. Perhaps the real impact of this is still a thousand or more years off, but sooner or later things will get very, very cold. And a great deal of adjusting will be necessary when that happens. The reason that there is so much concern with global warming is that right now the temperature appears to be going up at a dramatic rate, much faster than ever before. For my part, I agree that a catastrophe or several catastrophes are a real possibility.

Among the eight "imminent catastrophes" discussed in this book, I believe that the problems of global warming, the AIDS epidemic, and overpopulation deserve the most attention. And, indeed, global warming is receiving much attention. Many people strongly support the proposed measures to conserve energy, to reduce the emission of greenhouse gases, and to stop the decimation of rain forests. They do so primarily because they view them as ways to combat and to counteract trends in our technology-driven society that they feel are rapidly diminishing the quality of their lives: overcrowding, urbanization, pollution, destruction of treasures of nature, etc. So do I.

The measures now being advocated to reduce global warming may entail great expense and inflict economic hardships on many people. In addition to assuring a healthier environment, the monies spent in concerted efforts to control the amount of energy consumed around the world and to minimize the emission of greenhouse gases can be viewed as insurance—insurance against the possibility that the gloomy predictions of the global climate modelers are correct and that a global warming catastrophe might actually be imminent. Buying the insurance is a prudent course recommended by many in that Community. I agree.

Yet, I believe that prudence also demands that we prepare for the possibility that global warming will come upon us because the ambitious measures recommended to stop it will fail. The population of the world will continue to grow, people will continue to reach for higher standards of living, and, in the process, consume increasingly higher amounts of energy. Hence, if the scientific models are correct, or only partially correct, serious dislocations become almost inevitable. If we know that the seas will rise, we should begin to prepare for that eventuality by pulling settlements back from exposed coastal regions, by constructing dikes and breakwaters. We should begin now to reorient our agriculture so as to better adapt to the expected changes in climate. We should begin to act as if we really believe that disastrous global warming will occur in the coming decades. To do so is not to acknowledge defeat in the struggle against global warming. It is a prudent investment in another form of insurance.

POSTSCRIPT: THE ICE AGE IS COMING

> *THE ICE AGE IS COMING, within five to seven years from 1988.*
> —EPHRON[7]

While a large and influential body of scientists has been raising storm signals to warn of an impending catastrophe due to global warming, a small but passionate contingent of scientists and their supporters has discerned the possibility of a catastrophe due the lack of warmth. Both groups of scientists use the same basic data and observations, but they reach diametrically opposite conclusions. A large majority maintains that the observed increase in atmospheric CO_2 has already caused an alarming rise in temperatures around the world and will lead to further catastrophic heating. The other group interprets the data to show that temperatures have been declining for some 120 years and that we may already have entered a new ice age that will shortly lead to catastrophic cooling and freezing. Now, it is very likely that in the course of the next one thousand years, our beleaguered planet will experience extreme heat as well as extreme cold, so that both sides could be correct in the long run. However, here we are talking about the short run—predictions of global temperatures for the immediate future, the next twenty years or so. Both predictions cannot possibly be right.

The fact that this postscript comes immediately after the section on global warming should not be taken as an indication that the two contradictory theories enjoy equal support among scientists. In fact, an overwhelming majority of professional climatologists strongly favors the warming theory, at least for the short term. The ice age theory is included here not to "give equal time" to competing ideas, rather to highlight the difficulty, perhaps impossibility, of conclusively validating or invalidating a complex theory on the basis of observations and measurements. In fact, until about twenty years ago, the "ice age" theory had the support of the climatology establishment and was widely publicized; it was a *Newsweek* cover story, among others.

The case for the ice age is presented by Dr. Larry Ephron[8] in his 1988 book, *The End: The Imminent Ice Age & How We Can Stop It*. Ephron surveys the work of many investigators and theorists, but bases his conclusions principally on the work of John D. Hamaker, one of the most pessimistic champions of the ice age scenario. David E. Fisher[9] published *Fire & Ice* in 1990 and presents a serious but less apocalyptic view, one that is a bit closer to mainstream science. Jon Erickson,[10] in his *Ice Ages: Past and Future*, argues that "perhaps global warming with man's assistance, might be able to hold off the next ice age for a little while longer."

THEORY

Major ice ages are now well known to have occurred in approximate 100,000-year cycles. Each ice age lasts about 90,000 years followed by a warm interglacial period, about 10,000 years in length. It is now around 10,800 years since the last ice age—so we are due. The generally accepted explanation of this cyclic pattern was offered by Milutin Milankovitch, who attributes it to slight changes or wobbles in the orbit of the earth around the sun. Hamaker has another idea.

According to Hamaker and his cohorts, as an ice age comes to an end and the glaciers recede, plant life finds rich and fertile soils and flourishes everywhere and dense forests spread over much of the earth's land surfaces. These trees serve very effectively to maintain a low level of carbon dioxide in the atmosphere. As time goes on, the minerals in the soil, which are essential to healthy plant life, are gradually leached and eroded. During the millennia that constitute the interglacial periods, the soils become gradually less fertile, which leads to a gradual "die-back," a dying off of trees all over the world. Since trees now convert less and less CO_2 into oxygen, smaller amounts of carbon dioxide are removed from the atmosphere, which leads to a steady increase of atmospheric carbon dioxide and greenhouse warming.

The warming in the temperate and tropical latitudes then promotes an increased evaporation of water in those regions. This leads to increased temperature differences between the equatorial and polar regions, which in turn produces greater storms and massive air movement. Large quantities of moisture are consequently moved toward the poles, producing increased cloudiness and greatly increased snow and ice in the temperate zones and near the poles. Since snow and ice reflect light and heat, the temperature near the poles goes down. This in turn produces glaciers that slowly expand toward the equator. As the glaciers move over the surface of the earth, they grind up large quantities of minerals. These pulverized minerals are distributed all over the earth by winds and water. After about 100,000 years, the land surfaces have been remineralized and are again fertile. A new interglacial period with gradually increasing warming then commences.

Note a key difference between the warming and the cooling theories. The greenhouse effect partisans talk primarily about the average temperatures all over the world. By contrast, the ice age people consider *differential temperature effects*, the differences in temperatures and atmospheric carbon dioxide between the low and the high latitudes.

DATA

Proponents of the ice age theory cite vast amounts of empiric information, observational data to buttress their case. They are in general agreement with the greenhouse-warming theorists in observing a buildup of atmospheric CO_2 over the past several centuries, but particularly during the past hundred years where CO_2 levels have risen by almost 40%. The two camps differ markedly, however, when it comes to inferences regarding temperature trends. The basic problem is that annual changes in temperatures (and secondary effects, such as changes in sea level) are so small that they are almost completely masked by seasonal and other trends that are not related to long-range climactic changes.

Whereas supporters of the warming theory cite data showing a small but steady increase in overall (average over the entire globe) temperature in the past one or two decades, ice age theorists point to decreasing temperatures at the higher latitudes. Some observations made over the years in a number of locations all over the world lend support, at least in the eyes of some, to Hamaker's theory of a differential greenhouse effect that leads to warming in the tropics but to refrigeration in the higher latitudes. Other observations are cited to suggest that windstorms and earthquakes have, in fact, increased substantially—phenomena that are considered to be linked to increasing temperature contrasts between low and high latitudes.

A large number of observations are also offered to support the contention that the number of healthy trees around the world is declining rapidly. This is attributed only partly to human activities such as the destruction of tropical rain forests and air pollution. Increasing numbers of tree deaths are attributed to diseases (blights), insects, and naturally occurring forest fires. All of these are taken as evidence that soil demineralization is making trees less resilient and more likely to succumb to a variety of hazards.

PREDICTIONS

There are many scientists who accept all or parts of the ice age scenario. Ephron's book leans heavily on the theories of one of the most pessimistic of these—John Hamaker. According to him:

- The next ice age is overdue.
- The transitional stage of increasingly cold climate has already begun.
- We are very rapidly moving toward ice age conditions of extreme cold.
- By 1995 there could be massive worldwide food shortages and starvation. As many as half the people on the earth may die of hunger in the next five or ten years.
- There is a possibility of nuclear war because of dwindling resources.
- The next ice age cannot be stopped. It can only be delayed.

Several natural phenomena will act as agents of positive feedback and will hasten the onset of the ice age. Ice will continue to build up in the polar regions, and glaciers will continue to advance. This will produce an increasing number of volcanic eruptions, which in turn will increase the amount of polluting chemicals in the atmosphere, killing more trees and enhancing the greenhouse effect. At the same time, lasting sheets of snow and ice will cover more and more of the globe—reflecting more light and thereby promoting increased cooling.

According to Hamaker and Ephron, the future looks very bleak indeed. We must push a panic button **NOW** to delay the onset of the ice age as much as possible and to prepare ourselves for the worst.

WHAT MUST BE DONE

Some of the prescriptions of the ice age cohort are identical to those of the warming theorists. Both parties urgently call for a sharp reduction in the amount of CO_2 spewed into the atmosphere by human activities coupled with an increase in the flora which convert CO_2 into oxygen. Advocates of both the warming and the cooling theories agree that we should consume less energy in general and fossil fuels in particular, reduce atmospheric pollution and stop excessive logging activities.

But the ice age proponents go much further. They see trees as crucial in determining how quickly a catastrophic ice age will be upon us. Since forests all over the world are dying because of the low mineral content of soils, we have not a moment to lose to:

- Remineralize the forests of the earth.
- Plant billions of new fast-growing trees.

The effort and funds required to do this are mind-boggling. According to Ephron,[11] "We have to begin by spreading 3 tons of finely ground gravel dust on almost every acre of some 9 million square miles of forests and jungles: 5.7 billion acres, 17 billion tons of dust. . . . And more dust on millions of square miles where forests *could* grow, and then plant them. Wetlands,—marshes, bayous, and swamps—should also be remineralized. . . . Then we have to go back and do it again and again, to the limit of our vision and will, until we have spread enough gravel dust to completely rejuvenate the world's vegetation."

How can we do this? It is recommended that we:

- Stop all construction for perhaps a year so that cement grinders can be devoted exclusively to produce gravel dust.
- Ration air travel for two or three years and use all military and civilian aircraft to transport the gravel dust.
- Mobilize governments and volunteers all over the world to begin distributing the gravel, using thousands of trucks and other vehicles as well as newly developed blowers.

How much will this cost? According to Ephron,[12] "Probably on the order of a trillion dollars or so, give or take a few billion." But as he points out, that's about what the world spends on weapons and military activities each year. And society cannot afford to wait.

POLITICS

Proponents of the ice age theory are extremely frustrated to find themselves stymied at every turn by what they perceive to be a worldwide conspiracy. According to Ephron,[13] "It seems pretty clear that the United States government has successfully perpetrated a massive cover-up, actively suppressing a growing scientific consensus during the 1970s that the earth was cooling . . . and lending all its weight to the idea that CO_2 will inevitably cause the earth's climate to warm." He asserts that the U.S. government is a tool of the powerful energy industries (e.g., oil, coal, natural gas). These conglomerates oppose any phasing out or limiting of fossil fuels or of reducing the consumption of energy.

Ephron feels that the plot includes the funding of research to support the warming theory. Warming theorists are much more modest in their demands and are thinking in terms of the next fifty to one hundred years. By contrast, ice age theorists call for gigantic scale expenditures now. The scientists belonging to the warming Community have become the unwitting allies of the "military-industrial complex."

There can be no question that the bulk of the climatology Community supports the warming rather than the cooling theory. Some scientists lend lukewarm support to the general goals of the ice age activists. Many other leading climatologists are much harsher in their judgment of the ice age theory. They are quoted by Ephron as characterizing the ice age theory as "primitive," "dangerous," "ridiculous," and "absurd." This leads Ephron to observe that the most powerful members of the scientific establishment are ranged against him and his allies, that they refuse to accept important scientific papers in their prestigious journals and that they stand in the way of the necessary financial and political support.

Whatever the reasons, the ice age theorists have been far less successful than the greenhouse-effect Community in marshaling acceptance of their plans of action in the corridors of power as well as by the public at large. The prevailing sentiment is that the next ice age is hundreds, or perhaps thousands, of years in the future and that global warming is the imminent catastrophe that should concern us.

There are a number of additional recent books[14-28] that also address the problems of climate modification.

SEVEN

Nuclear Radiation

There is now reason to fear that low-level radiation from fallout and from nuclear reactors may have done far more damage to humans and other living things than previously thought, and that continued operation of civilian and military nuclear reactors may do irreversible harm to future generations as well.

—J. M. GOULD AND B. A. GOLDMAN[1]

Visions of the mushroom-shaped clouds over Hiroshima and Nagasaki are permanently etched in our collective consciousness. Indeed, in the minds of most people, at least until the recent thaw in the relations between the superpowers, the threat of nuclear war ranked as the most terrifying of all catastrophe scenarios.

The explosion of a nuclear bomb gives rise to intense heat and to shock waves that incinerate or pulverize everything in the immediate vicinity. More dreaded, however, is the effect of nuclear radiation on the health of all living things within hundreds of miles of the blast. The immediate and long-term results of large doses of radiation include severe burns, blindness, cancer, and other diseases as well as genetic damage that affects the yet unborn. An all-out nuclear war could unleash catastrophic effects on the climate—the "nuclear winter" discussed by Carl Sagan and Richard Turco[2] among others. As of 1990, however, no scientist was prophesying a nuclear war and nuclear holocaust. Instead, attention shifted to a less dramatic but much more imminent nuclear threat—low-level radiation.

For many years after World War II, many scientists and a large segment of the public believed that, while intense radiation is undoubtedly harmful, there is little to fear from low levels of radiation. After all, we are all constantly exposed to radiation from outer space, from chemicals occurring naturally in the earth, and from a wide variety of technological devices, yet no ill effects are evident. This view was reflected in the policies of most governments in pursuing military and industrial applications of nuclear energy.

Prominent scientists, such as Linus Pauling in the United States and Andrei Sakarov in the Soviet Union, raised early storm warnings in connection with the testing of nuclear bombs in the atmosphere. They and their colleagues waged a long and intense campaign to alert the world to the dangers of airborne radioactive particles. By the early 1960s, the detrimental effects of radioactive fallout from these tests became so well recognized that, in 1963, all such experiments were suspended.

The potentially dangerous side effects of peaceful uses of nuclear energy then became the focus of concerns, and nuclear reactors, particularly the large power reactors, became the targets of environmentalists worldwide. The big fear was of accidental explosions of such reactors, explosions very much like nuclear bomb blasts. Soon, small accidental emissions from reactors, as well as the accumulation of radioactive waste products, also came under fire for constituting significant hazards. Given the increasing demands for electrical energy, particularly in the developing countries, and in the absence of practical alternative sources of energy, decision makers all over the world have tried and continue to try to select the lesser of two evils: nuclear power with the attendant radiation or fossil fuel plants that emit dangerous air pollutants and promote the greenhouse effect.

The disastrous nuclear accident at Chernobyl in April 1986 and its aftermath aroused public concern to a fever pitch, motivating efforts to reexamine the dangers of nuclear power. Yet, long before this incident, a number of prominent scientists predicted long-term health problems of catastrophic proportions, problems caused or exacerbated by levels of radiation heretofore considered of little consequence.

In 1990, in their book, *Deadly Deceit: Low Level Radiation, High Level Cover-up*, Jay M. Gould and Benjamin A. Goldman[3] present the results of a very comprehensive and detailed statistical study of the correlation between known or suspected emissions of radioactive particles into the atmosphere and health problems ranging from diminished intelligence, to increased infant mortality, to the incidence of AIDS and other dreaded diseases. These data gave support to many of the conclusions published in 1988 by Robert P. Gale and Thomas Hauser[4] in their *Final Warning: The Legacy of Chernobyl*. Derek Elsom's[5] *Atmospheric Pollution* provides some interesting background information regarding the threat of radioactive matter.

It should be noted that many of the interpretations of data and many of the conclusions presented in the books mentioned above are opposed and contradicted by scientists belonging to the Community advocating nuclear power. For example, Bernard L. Cohen,[6] in his recent book *The Nuclear Energy Option: An Alternative for the 90s*, presents arguments implying that the threat of nuclear radiation has been greatly exaggerated, that super-safe nuclear reactors will soon become available, and that the safe disposal of nuclear wastes is well within our present means.

THEORIES

In the two decades following World War II, the United States and the Soviet Union conducted frequent atmospheric tests of nuclear weapons. The estimate is that during those twenty years they discharged a volume of radioactive debris into the atmosphere equivalent to that released by over 40,000 bombs of the type that destroyed Hiroshima. During that time, relatively little care was taken in limiting accidental emissions from nuclear power plants—and there were some substantial emissions. Moreover, according to Gould and Goldman,[3] there have been many unreported incidents since then, and, of course, then came Chernobyl. It is now the conviction of many scientists that the occasional release of radioactive particles (actually gas and dust contaminated by radioactive isotopes) is an almost inevitable consequence of the operation of reactors that work by nuclear fission. Fusion reactors that do not produce radioactive matter might be an attractive alternative, but they are still on the drawing board and will not become available until the next century, if at all. Hence, we have been bombarded by radiation for over forty years, and we will continue to be so as long as we rely on nuclear energy.

The damaging effects on individual and public health of high doses of radiation have been determined more or less conclusively by exhaustive studies of the victims and the survivors of the bombings of Hiroshima and Nagasaki in 1945 as well as of a number of accidents since that time. By contrast, it is very difficult to come to grips with the insidious long-term consequences of relatively small amounts of radiation because of uncertainties as to the causal chains that link these radiations to eventual illnesses or mortality. There appears to be a number of possibly significant mechanisms.

Radioactive waste is formed in the nuclear fission processes that take place in nuclear reactors as well as in bombs. Most of this waste is in the form of radioactive isotopes of relatively common chemical elements. The isotopes of an element all have the same number of protons in their nucleus, but different numbers of neutrons. Radioactive isotopes are unstable; they decay by changing into other isotopes of the same element or into an entirely different element. Sometimes this decay takes fractions of a second; in other cases it may take many years. For example, radioactive iodine-131 has the same number of protons as the stable iodine-127, but it has four more neutrons. It decays by emitting high-energy electrons called *beta particles* and by changing into radioactive xenon-131. Eight days after it is formed, half of the iodine-131 will have disappeared. We say that it has an eight-day *half-life*. The xenon-131 decays in turn, with a twelve-day half-life. Similarly, strontium-88, strontium-89, and strontium-90 are all isotopes of the common metal strontium. But only strontium-88 is stable. Radioactive strontium-89 has a half-life of 54 days, while that of strontium-90 is 28 years. That makes strontium-90 a very long-term hazard. Table 5 shows a number of often-encountered by-products of nuclear reactions and their half-lives.

Because they are very small and light, radioactive gas and dust particles rise easily once they are released into the atmosphere, and they are quickly distributed all over the world by winds and air streams. Some eventually settle down to earth just like other dust particles. Many of them have an affinity for moisture and are returned to earth dissolved in rain water. In fact, while some radioactive particles may be inhaled, as much as 90% of the radiation damage is done by airborne particles that come down with the rain.

All of the particles listed in Table 5 are harmful and capable of causing serious health damage. It is generally agreed that the real killers are iodine-131 and strontium-90, both of which are formed from the iodine and strontium that make up a very small part of the fuel of nuclear reactors (and bombs). In fact, during the turbulent years of World War II it was seriously suggested that atom bombs laced with strontium be used to poison the German food supply.

This is how iodine-131 and strontium-90 (and other radioactive particles) are thought to enter the human body. Radioactive particles are released into the atmosphere in the course of nuclear reactor accidents or as a result of improper waste disposal. They are dispersed in the air over large areas and then descend dissolved in rain drops. The rain water is absorbed by plants, such as grass, and eventually consumed by herbivorous animals, including important sources of food such as cows, sheep, goats, etc. When the milk or the meat of these animals is consumed, the

TABLE 5. Some of the Many Sources of Radioactive Particles
(from Elsom,[5] by permission of Blackwell Publishers)

Radionuclide[a]	Half-life	Principal radiations		
		Alpha particles	Beta particles	Gamma rays
Argon-41	1.8 hours		✔	✔
Barium-140	12.8 days		✔	✔
Cesium-134	2.1 years		✔	✔
Cesium-137	30.1 years		✔	
Cesium-139	9.5 months		✔	✔
Carbon-14	5.7×10^3 years		✔	
Cerium-141	33.0 days		✔	✔
Cerium-144	284.0 days		✔	✔
Iodine-129	1.6×10^7 years		✔	✔
Iodine-131	8.1 days		✔	✔
Iodine-132	2.3 hours		✔	✔
Krypton-85	10.7 years		✔	✔
Molybdenum-99	1.1 days		✔	✔
Plutonium-237	46.0 days			✔
Plutonium-238	87.8 years	✔		
Plutonium-239	2.4×10^4 years	✔		
Plutonium-240	6.5×10^3 years		✔	
Polonium-210	138.0 days	✔		
Potassium-40	1.3×10^9 years		✔	✔
Radium-224	3.6 days	✔		✔
Radium-226	1.6×10^3 years	✔		✔
Radium-228	5.8 years		✔	
Radon-222	3.8 days	✔		
Ruthenium-103	39.5 days		✔	✔
Ruthenium-106	1.0 years		✔	
Strontium-89	50.5 days		✔	
Strontium-90	28.5 years		✔	
Tellurium-132	1.3 hours		✔	✔
Uranium-233	1.6×10^5 years	✔		
Uranium-234	2.5×10^5 years	✔		
Uranium-235	7.1×10^8 years	✔		✔
Uranium-236	2.4×10^7 years	✔		
Uranium-238	4.5×10^9 years	✔		
Uranium-239	23.5 months		✔	✔
Xenon-133	5.3 days		✔	✔
Xenon-135	9.2 hours		✔	✔

[a]Isotopes are characterized by a number that represents the total number of particles (protons plus neutrons) in their nuclei, e.g., uranium-235 has 92 protons and 143 neutrons whereas uranium-238 has 92 protons and 146 neutrons.

radioactive particles are digested, enter the blood stream, and eventually go their separate ways.

Iodine-131 travels straight to the thyroid gland, one of the endocrine glands that control our metabolism and many bodily processes throughout our lives. The chemical, even in tiny concentrations, is especially harmful to fetuses and to infants since the thyroid gland controls the growth hormones. It is theorized, therefore, that iodine-131 is responsible for miscarriages, immature and underweight babies, and many cases of infant mortality as well as the deficient physical and mental development of some of the infants who survive. A number of diseases, including various forms of cancer, have also been attributed to iodine-131.

Strontium-90 has a much longer half-life and is therefore capable of accumulating in the human body over many years. This radioactive isotope appears gradually to find its way into the bones of its victims, where its concentrations typically reach a peak some three years after it enters the body. There are many theories regarding the manner in which it can then affect physiological processes. Strontium-90 has been linked to cancer, heart attacks, miscarriages, and many other fatal health problems. Some researchers believe that it acts directly to damage the human immune system, making its victims more susceptible to various kinds of infections. Pregnant mothers, fetuses, infants, and old people are thought to be particularly at risk. The susceptibility of people born in the nuclear age to immune-system-related diseases (including the Epstein–Barr virus, Lyme disease, herpes, and especially AIDS) has been used to explain, at least in part, the rising incidence of these modern plagues (see Ref. 8).

The presence of radioactive matter in the body is also believed to promote mutations and thereby the rise of hostile organisms. In fact, one highly controversial theory links the origin of the AIDS virus to the atmospheric nuclear tests of the 1950s and early 1960s, which were conducted at the latitudes of the heavy rainfall areas of Central Africa.

As already pointed out, scientists have known for a long time that large doses of radiation are harmful. The long-term effects of low-level radiation and the mechanisms of these effects continue to be the subject of controversy. Some people even believe that small amounts of radiation are good for you, the cure of many ailments, as witnessed by the numerous hot spring resorts that feature radioactive waters. Some comfort in this regard is offered by detailed analyses of the victims of Hiroshima and Nagasaki. The degree of health damage to these individuals is closely related to the extent of their exposure to radiation. These figures suggest that very low levels of radiation are not harmful.

It has been theorized, however, that biochemical processes, quite different from those that attend exposure to high levels of radiation, may become active and dominant at very low radiation levels. For example, the Canadian biologist Abram Petkau has proposed a mechanism whereby radioactive products accumulating in the human body may damage the immune system. The so-called "Petkau effect" entails the formation of free radicals, such as a special form of oxygen, that attack cell membranes, destroying the cells or causing undesirable mutations and a number of other kinds of physiological problems. This effect takes place only when radiation levels are very low and is apparently suppressed by higher radiation levels. During the past twenty years, Petkau and his colleagues have carried out many experiments and published a long list of papers on the subject. Others have shown that free radicals may cause heart disease, cancer, and other maladies. All this has established the Petkau effect at least as a candidate link in a plausible causal chain. According to these still-controversial theories, radiation is harmful no matter how small the dose.

This view is supported by recent statistical studies showing a strong correlation between very low radiation levels in the vicinity of accidental emissions and a variety of health problems, but especially cancers and birth defects. If these theories become generally accepted, it may be necessary to revise drastically the manner in which we deal with all sources of radiation—not only radioactivity, but also very low levels of X rays, radon, and electric fields.

DATA

The tenuous links of low-level radiation to human health problems are based on isolated laboratory studies of various biochemical processes and on statistical analyses of the incidence of a variety of ailments.

Gould and Goldman[9] have made careful statistical studies of data obtained from official death certificates in various parts of the United States, information that is open to the public. They compared mortality data over an extended period with data collected immediately after nuclear reactor accidents. In this way they are able to determine the increase in mortality that was presumably caused by radiation emitted in the course of the accidents. Additionally, they analyze changes in iodine-131 and in strontium-90 in milk and other food supplies and in rain water. The following are examples of their findings.

Data show that infant mortality in the southeastern part of the United States increased substantially immediately after the Chernobyl accident. This rise appears to be closely correlated with the observed increases in the iodine-131 content of milk in that geographical area. This is closely correlated with the amount of radioactive iodine-131 in fresh farm milk. A similar effect was observed at that time in West Germany.

Other data show the monthly changes in infant mortality as well as in total mortality in South Carolina and in the entire southeastern United States for a period during which a series of accidents occurred at the Savannah River nuclear weapons plant in South Carolina. Similar graphs are presented for the Three Mile Island accident and for incidents during the start-up of the Peachbottom reactor in Pennsylvania. Other data establish correlations between the incidence of cancer in infants, birth defects, and other ailments as well as nuclear testing and accidents.

Such statistical information cannot be regarded as proof positive of a causal relationship between low-level radiation and sharp increases in the number of deaths. It is possible, nonetheless, with the aid of statistical tools to assign a probability to the possibility that observed correlations were due to chance rather than that they reflect some kind of causal relationship, and for the data presented that probability is very small indeed—of the order of one in one million in some instances.

THE CAUSAL CHAIN

The overall model for the prediction of catastrophes as a result of low-level nuclear radiation has the three major components shown in Figure 24. First, we have the sources that emit radioactive matter into the atmosphere. The radioactive air pollutants are due to accidents in nuclear power plants, the unsafe disposal of radioactive waste products, as well as other causes, all encompassed by Model A. These then make their way into the human and animal food chain via water, grass, and so forth. The output of Model B is the amount of radioactive matter ingested by people and by animals. Model C predicts the extent to which this radioactivity causes serious health problems, such as cancer and birth defects.

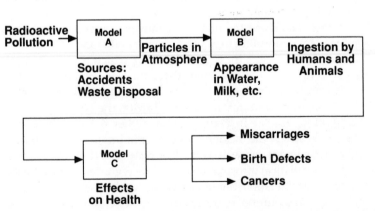

FIGURE 24. Comprehensive model of the threat from low-level nuclear radiation.

In this comprehensive model, the output of Model A is a big question mark. For whatever reasons, governments and utility companies in past years have not made public the amount of radioactive matter that they have released into the environment. Why this information is not available is beside the point; the point is that no one knows, even approximately, what has been emitted into the air, into ground water, and into the ocean. Nor does anyone know how many poorly engineered dumps of radioactive waste are secreted all over the world, sooner or later to leak their contents into the environment. Therefore, in terms of the Spectrum of Models of Chapter 3, Model A has many dark gray features.

Model B is also of the dark gray variety. Radioactive particles emitted into the atmosphere, as at Chernobyl, are dispersed by winds and may travel for hundreds and hundreds of miles before descending to earth along with rain. Since we have great difficulties in forecasting the weather, the prediction of the location and intensity of radioactive fallout is also subject to great uncertainties. The effect upon the public health of ingesting radioactive matter, as mapped by Model C, is reasonably well understood, at least in statistical terms. The incidence of cancer, miscarriages, birth defects, and other serious consequences can be generally correlated with increased levels of radioactivity in milk and water. But quantitative predictions of future health problems on the basis of the comprehensive model belong in the very dark region of the spectrum.

PREDICTIONS

The outlook for the future is very gloomy indeed. Over the past thirty-five years, enormous quantities of radioactive materials, some with very long half-lives, have been released into the atmosphere, the oceans, lakes, streams, and underground water reservoirs. The atmospheric nuclear bomb testing that ended in 1963 was just the beginning. In addition to the known nuclear reactor accidents, there may have been a number of incidents that were hushed up, at least according to some observers.[3] Then there are the radioactive wastes, many improperly stored. The environment has been terribly mistreated for a long time, and there is little that can be done about it now.

Projecting from the past into the future, we must expect a persistent increase in excess deaths, deaths above those to be expected in the absence of the radiation phenomenon. Gould and Goldman[3] present a comparison of the mortality rates that would be predicted when projecting pre-1950 data (showing a progressively increasing life expectancy) with predictions based upon the observed excess deaths since 1950. By 1985 the cumulative total of these excess deaths in the United States had reached nine million. If present trends continue, the average expected life span of all people in the United States would eventually begin to decrease, canceling out all potential gains due to advances in medicine and improvements in the standard of living. To this already bleak picture must be added the possible effects of AIDS as well as other as yet unrecognized diseases, allegedly caused by radiation-related damage to the immune system, and the emergence of newly evolved viruses.

WHAT MUST BE DONE

Many of the results of our past sins involving low-level radiation are irreversible. All that we can hope to do is refrain from making things even worse. The Community says we must convince our governments to phase out and eventually close all nuclear facilities throughout the world.

The U.S. government has estimated the cost of getting rid of its military nuclear plants: $175 billion. The cost for doing this for all civilian nuclear facilities may well exceed several thousand billion dollars. And that's for the United States alone. Remember that many countries are counting on nuclear energy to help them to develop their economies and pull their standard of living closer to that of the developed countries. Also, major research and development efforts are required to devise means for the relatively safe disposal of the radioactive wastes that have accumulated and that will continue to increase in volume until all reactors are turned off.

POLITICS

Many countries have had stakes in the nuclear energy field, both military and civilian. It is now clear that the extent of environmental pollution and the effect of low-level radiation on people living in the vicinity of nuclear activities have been minimized or concealed by the governments of the United States and the Soviet Union. Will this change in the current climate of reduced international tensions?

A PERSONAL VIEW

The damage inflicted by nuclear reactor mishaps in the past is manifest. The release of harmful radioactive materials into the atmosphere and into water supplies by future reactor accidents and by the imperfect disposal of nuclear waste products constitutes a real danger. The careful and continuous monitoring of radiation levels all over the world is essential, and every effort must be made to eliminate the misinformation, the seeming cover-ups relative to accidental emissions that have accompanied past reactor accidents. But should we completely write off the generation of electric power by nuclear reactor facilities? Should we curtail our efforts to develop and to implement safer nuclear reactors? I believe not.

I fully approve of current efforts to reduce the consumption of energy by increasing the efficiency of power hungry machinery, and I support the development of "alternative" energy sources. But I fear that these will be only partially successful at best. The population of the world will continue to grow and people will continue to require energy. Even if regional efforts to control the population meet with success, as appears to be the case in China, the demand for energy will not diminish. Past experience shows that as a population explosion abates, the standard of living increases. And improved standards of living are accompanied by demands for more appliances, expanded transportation systems, and more elaborate facilities of all kinds. For instance, China is now striving to provide a refrigerator for every family. Air conditioning may be next. Hence, the consumption of electricity by more than one billion Chinese will go up and up. And what about India, Indonesia, and, eventually, Africa? Let's face it, the world will be consuming far more energy in the future than it does now.

And how is this energy to be generated? Fossil fuels are the primary source at present. As demand increases, so will the use of gas, oil, and especially coal. This will bring us ever closer to catastrophic air pollution, global warming, and a host of other tribulations. I believe that there will come a point at which nuclear power, even with all of the attendant radiation problems, will be recognized as constituting a lesser evil. Perhaps nuclear fusion will come to the rescue. More probably, though, we will still have to work with nuclear fission, and we must hope that the large and hazardous power reactors now in use will have evolved into more manageable and safer models.

EIGHT

Air Pollution
Acid Rain

Acid rain is generally regarded as one of the most serious environmental problems of our times. . . . It ranks alongside important contemporary concerns like the global increase of carbon dioxide in the atmosphere and the possible environmental consequences of nuclear war.

—Chris C. Park[1]

The problem of acid rain is similar to that of low-level radiation in that its disastrous effects are already with us, having gradually worsened over many years; but the real catastrophe is still ahead. While there are serious health problems associated with acid rain, its most direct consequences touch animal and plant life more than people. These are already serious enough to have given rise to intense international activity on scientific as well as political levels.

Acid rain is primarily the result of the emission into the air of a number of polluting agents, all of which act to increase the acidity not only of rain, but also of snow, mist, and other forms of

179

airborne moisture. The term is also applied to pollutants that are deposited on the earth in dry form and which do not become acidic until they come in contact with water. While unpolluted rain water is always acidic to some extent, certain pollutants effect an increase in this acidity to a level that makes acid rain damaging in many ways.

Acid rain is toxic to aquatic as well as terrestrial animal and plant life. The increased acidity of lakes, streams, and other bodies of water is prejudicial to the welfare of fish, and the population of many species has declined sharply; some species have completely disappeared. On land, acid rain is blamed for the wholesale death of trees, even entire forests, in many parts of the world. There is also evidence that acid rain is responsible for the increase of certain toxic chemicals in drinking water and that it is an indirect cause of a number of diseases.

Chris C. Park's[2] book *Acid Rain: Rhetoric and Reality*, which first appeared in 1987, and Gwyneth Howells's[3] *Acid Rain and Acid Waters*, which was published in 1990, provide excellent introductions to the subject. David D. Kemp's[4] *Global Environmental Issues: A Climatological Approach* and Derek Elsom's[5] *Atmospheric Pollution* place the subject in a broader context.

THEORIES

Acid rain is formed when pollutants containing oxides of sulfur and nitrogen are emitted into the atmosphere. These gases diffuse in the air and can be transported by winds and air currents as far as hundreds of miles.

In the presence of sunlight and of atmospheric moisture, these pollutants are oxidized and form acids, such as sulfuric acid, H_2SO_4, and nitric acid, HNO_3. In rain clouds, these acids are turned into charged particles or ions, such as radicals of hydrogen and oxides of sulfur and nitrogen, which eventually descend to earth where they begin to do their mischief.

Chemists measure the strength of acids using the logarithmic "potential hydrogen" or pH scale. Distilled water, which is neutral, has a pH level of 7.0 and the very strongest acids have pH levels near 0.0. Milk has a pH level around 6.7, orange juice 3.3, and vinegar 3.0. Because of the logarithmic nature of the pH scale, a chemical with a pH level of 4.0 is ten times as acidic as one with a pH level of 5.0. The pH level of unpolluted rain water is usually in the vicinity of 5.6, an acidity that it picks up from carbon dioxide and other naturally occurring components in the atmosphere. By contrast, typical acid rain has acidities ranging from under 4.0 to 5.0. Rain with a pH level below 4.5 is generally considered to be harmful.

The problem is complicated by the fact that the atmospheric chemical reactions that produce acid rain are affected by a number of chemicals. It is generally accepted, however, that the principal culprits are sulfur dioxide and various oxides of nitrogen identified by the catchall term NO_x. Physical variables that play a part include the intensity of the sunlight, the kind of cloud cover, and the wind turbulence to mention just a few. There is less agreement as to the major sources of these pollutants and how they damage the environment.

A considerable amount of sulfur dioxide and oxides of nitrogen are generated continuously by natural processes. Oceans produce and emit much sulfur dioxide, while many biological processes generate oxides of nitrogen. Since it is very difficult to estimate the total amount of these pollutants that is formed naturally, it is also very difficult to determine the excess amount for which human activities should be blamed. Still, there is no shortage of theories.

As far as sulfur dioxide is concerned, the principal human sources of the pollutant are thought to be:

- Burning of coal to generate heat and electricity—60%
- Burning of petroleum products for heat and electricity—30%
- Smelting of metallic ores and other industrial processes—10%

Oxides of nitrogen, which are of much smaller but still appreciable importance, are produced by electric power stations, by the exhausts of automobiles and other vehicles, and by industrial processes. The exact breakdown is very much open to debate.

The generation of man-made acid rain pollutants varies much from place to place. The industrial regions of the United States, Europe, and Japan are certainly responsible for most of the oxides of sulfur and nitrogen emitted into the atmosphere by human activities. Precisely how much, though, depends on the model and is the subject of endless controversy. Likewise, there is little agreement as to the extent to which and the manner in which acid rain wreaks its havoc. Early reports of damage to flora and fauna came from southern Scandinavia, eastern Canada, and parts of western and central Europe, all of which are located downwind from major industrial complexes. There is now great concern that the entire globe will soon exhibit some of the effects of acid rain; symptoms, such as blighted plants, have already been observed in a number of Third World areas.

Biologists have made careful studies of the effect of water acidity on marine life. Many aquatic organisms are so sensitive to the acidity of their habitat that very few can survive when the pH drops below 5.5. Many theories have been offered to explain this phenomenon in terms of biochemical processes. Acid rain products are thought to be responsible not only for an increase in acidity but also for the appearance of toxic minerals, such as aluminum and cadmium, which are leached from the soil and dissolved in the polluted streams and lakes.

There are also numerous models to describe the effect of acid rain on vegetation. Small amounts of acid are known to be beneficial, even vital, to the health of most plants; it is high concentrations of acid that cause problems. Atmospheric sulfur dioxide is known to be directly injurious to plants that absorb the chemical through their leaves. This produces immediate damage to plant tissues which can eventually destroy the plant. It is thought to be a major cause of the tree deaths that have become widespread in Europe and eastern North America. A second problem is posed by the acidification of the soil in which plants grow. This can greatly reduce the fertility of the earth and its ability to support plant and animal life. The chemical processes involved are numerous and complex. Available models are incomplete and, as yet, of low reliability.

Even more tenuous and incomplete are the models that link acid rain with human health. Some obvious connections have been made between air pollution and various respiratory maladies, and there have been suggestions of other direct and indirect effects. It would appear, however, that the catastrophic events that are predicted to result from acid rain will affect humans indirectly by sharply reducing important resources that are essential for food and building materials.

DATA

Over the past thirty years, enormous amounts of data regarding air pollution and its effects have been gathered and digested. Still in the eyes of many, the case against power plants and other industrial sources of pollution remains circumstantial.

Figures show that all over Europe the pH of rain water dropped by as much as 1.0 in the 1960s and 1970s. Germany has been particularly concerned about the effect of acid rain on its forests and has coined the term "Waldsterben" or "forest death" for this phenomenon. As far as sources are concerned, published estimates[6] of the sulfur dioxide emissions in Europe during the 1970s show an increase ranging from 20% in Sweden to 63% in West Germany to 102% in France and to 117% in Finland. Copious data are also available to document the manner in which SO_2 moves from country to country. Other data show that approximately 50% of the oxides of sulfur and nitrogen polluting the atmosphere of Great Britain during the 1980s were emitted by power stations, with lesser contributions from automobiles, refineries, and domestic sources.[7]

THE CAUSAL CHAIN

The comprehensive model of the acid rain problem has three major components, similar to the low-level radiation model of the preceding chapter. As shown in Figure 25, the first submodel, Model A, predicts the quantity of chemical pollutants, principally sulfur and nitrogen dioxide, that are emitted into the atmosphere by various human activities. Model B serves to estimate the geographical distribution and the amount of the sulfur and nitrogen compounds that are deposited on the ground and on water surfaces. The possibly catastrophic effects of these deposits on humans, animals, and plants are determined by Model C.

More and more data regarding industrial and urban sources of pollutants, as characterized by Model A, are becoming available. The extent to which these will be active in the future depends upon the success of the various mitigating efforts now under way. Scrubbers on oil refineries and fossil fuel plants, as well as catalytic converters for automobile exhaust, can become increasingly significant in reducing emissions. But to what extent is an open question. Hence, in terms of the Spectrum of Models, Model A is medium gray. The validity of Model B, which predicts the dis-

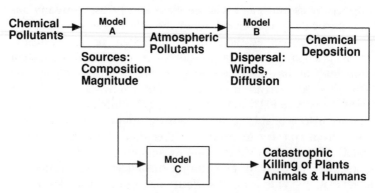

FIGURE 25. Comprehensive model of acid rain.

persal and eventual precipitation of the components of acid rain, is more or less the same as that for other meteorological models—medium gray. And Model C is a biological model and is therefore darker than medium gray.

PREDICTION

Many scientists predict the demise of most forests in those areas of Europe and North America that are already hard-hit by the effects of acid rain—this unless heroic measures are taken at once to abate the emission of the major pollutants. They further foresee a spread of the consequences of acid rain, until all regions of the globe are substantially affected.

WHAT SHOULD BE DONE

The Community exhorts us to stop emitting oxides of sulfur and nitrogen! We must repair the damage already done. Unfortunately, this is much easier said than done.

Approximately one-half of the pollutants that cause acid rain are emitted by power stations that burn coal or oil. There are a number of ways to reduce this amount.

1. Conserve energy so as to consume less fuel. A number of successful moves in this direction have already been undertaken in the developed countries. However, the energy requirements of Third World nations are expected to increase in the coming decades.
2. Switch to low-sulfur fuels. Some fossil fuels are much richer in sulfur than others. There are low-sulfur coals and low-sulfur oils on the market, but usually at a higher price than the garden variety.
3. Switch to alternative fuels. These include natural gas, wood, and kerosene as well as a number of so-called renewable energy sources, such as the sun, tidal forces,

hydroelectricity, and winds. Nuclear power would also help mitigate the acid rain problem, but perhaps this would eventually inflict even greater damage to the environment in the form of low-level radiation.

4. Clean coal before burning. There are methods of "washing" coal to remove about 10 to 15% of its sulfur.

5. Remove sulfur during or after burning. This can be done by mixing limestone with the coal. Alternatively, the gaseous combustion products can be "scrubbed" by passing them over or through limestone.

There are also a number of practical ways of limiting the emission of oxides of nitrogen by power stations and by internal combustion engines. These involve the addition of special exhaust scrubbers and the introduction of improved automobile engines and gasoline.

All of the above remedial measures entail considerable expense, both in installation and in operation. For example, the U.S. Office of Technology Assessment estimated that to reduce the level of atmospheric sulfur dioxide in the eastern United States by 35% by 1995 would cost anywhere from three to six billion dollars. For that reason, the adoption of draconian remedial measures has often been opposed, not only by the energy industries but also by a number of governmental agencies.

There is evidence that some irreversible damage has already been done to the environment. That is, even if the emission of all man-made acid rain pollutants were to cease at once, the environment would not return to its pre-acid-rain state. Too much acid in the soil and in the water already exists in many places. To counteract the presence of the acids that were deposited during acid rain episodes, it is necessary to "sweeten" the soil and water by introducing large quantities of lime. This method has been used with some success in Sweden, but at a very considerable cost. To employ this approach on a worldwide basis would entail enormous expenditures for the mining, the conditioning, the distribution, and the application of the powdered limestone.

POLITICS

Notwithstanding the huge costs involved, a number of governments have taken some halting steps in the right direction. In 1979, a United Nations commission drafted rather loosely worded guidelines entitled "Convention on the Long-Range Transportation of Air Pollutants." This was followed in 1985 by a legally binding protocol that required all signatories to reduce their sulfur dioxide emissions by 30% of their 1980 levels, and to do this by 1993. Neither the United States nor Great Britain signed this protocol, and both have been engaged in heated controversies with their neighbors. However, in late 1990, the U.S. Congress passed a bill mandating a ten-million-ton reduction in the emission of sulfur compounds. Later that year, the National Acid Precipitation Assessment Program, sponsored by the U.S. government, released a report summarizing ten years of study and modeling of the acid rain problem in the United States, an effort that cost over one-half billion dollars. That report caused a storm in the political arena and the media, because it appeared to down play the threat posed by acid rain.

A PERSONAL VIEW

As a resident of Los Angeles for the past forty years, I am only too familiar with the obnoxious and the noxious indications of air pollution. After all, we Los Angelenos invented smog. For several decades, scientists from all over the world have traveled to our fair city to study this phenomenon, a tourist attraction of sorts. Eventually, other cities around the globe caught up with us, and quite a few have surpassed us as smog generators, but we were the first to become famous for it.

From the 1940s on, the effects of Los Angeles air pollution have been painfully evident. There are myriad health problems: respiratory ailments, skin diseases, heart attacks, and even lung cancer; agriculture, especially the orange groves, suffer visibly;

industrial products, from rubber tires to nylon stockings, wear out much more rapidly; and, of course, for all to see is the brown layer of polluted air covering the L. A. basin and neighboring regions, and not infrequently seeping all the way to the far-off Grand Canyon. This is not exactly acid rain, since it rarely rains in Los Angeles, but very similar in many of its effects.

Which air pollutants were responsible for which symptoms remained a mystery for a long time. Government agencies, university laboratories, and private foundations struggled with the problem for over forty years.

The first suspected culprit was smoke from backyard incinerators. Los Angeles did not have a municipal trash collection service, and I still remember seeing the familiar plumes of smoke rising from each residence as I drove home from work each evening. In the early 1950s, Los Angeles outlawed the use of backyard incinerators, and fleets of trucks fanned out to pick up the rubbish. Unsavory political squabbles and charges of graft related to the award of lucrative contracts for this service ensued. Yet, ultimately, these were sorted out. However, the smog remained virtually unaffected. Attention then was focused on the oil refineries and some of the other major industries that were compelled to install expensive equipment to scrub and filter all gaseous efflux. That did make a difference. The concentration of atmospheric sulfur dioxide was definitely reduced, but most of the harmful effects of smog remained more or less unchanged.

Finally, in the 1960s, scientists identified ozone as the principal offender. In small concentrations, this close relative of oxygen is not unpleasant. But when it reaches a certain concentration level it becomes very unhealthy, even deadly. On smoggy days, the ozone concentration in Los Angeles often rises well above that threshold. Once it was agreed that ozone was at the root of the problem, the big question became: "How do you get rid of the ozone?" Unfortunately, there is no way to pipe it up to the stratosphere to help fill the growing hole in the ozone layer. It turns out that ozone is not released into the atmosphere in appreciable quantities by any known source. Rather, in the presence of sunlight, oxides of nitrogen, primarily from automobile exhaust, react with oxygen in the air to form ozone and other chemicals. This reaction is reversible; when the sun goes down, most of the ozone disappears, only to reappear again the next morning.

When this causal chain was finally understood, Southern California governments adopted a series of regulations mandating the installation of exhaust filters on most cars as well as regular inspections to make sure that all vehicles met specified exhaust standards. Has all this solved the air pollution problem? By no means! At least things have not gotten worse as the number of automobiles in the area has increased, and some statistics do suggest that some noticeable improvements have been effected. In 1989, California adopted the Clean Air Plan, which imposes progressively more stringent controls on all sources of air pollution—in industry, in business, and in the home. Over a twenty-year period, the plan is expected to cost about one hundred billion dollars. And that is not counting losses to businesses that become less competitive, jobs lost to other states and countries, and hardships imposed on consumers. This is, therefore, a continuing and very expensive battle.

The story of the fight against air pollution in Los Angeles may in some ways serve as a model for global projects to combat acid rain and other forms of chemical air pollution. Progress can be made through concerted actions, but only at a very high cost.

On my personal scale of imminent catastrophes, acid rain ranks rather low. It is not that I am indifferent to the problem. There can be no question that oxides of sulfur and nitrogen in the atmosphere are very unhealthy for humans, animals, and plants. However, the quantitative relationship between air-pollutant concentrations and the resulting damage to life forms is still open to controversy. Clearly, the reduction of oxides of nitrogen and of sulfur in the exhausts of industrial plants, especially power stations, and of motor vehicles is an important and worthwhile goal. But, in my opinion, the probable effects of undiminished emissions of these pollutants fall short of constituting an imminent global catastrophe. This is true also of the cost/benefit trade-offs inherent in various proposed strategies to abate the emission of acid rain chemicals.

The high cost of disaster mitigation is a problem common to all of the catastrophes discussed in this book. In each case, there is a running battle that has pitted a science-based advocacy Community against the industries, businesses, and taxpayer advocates that object to the high price of the proposed mitigation. In the case of air pollution controls, this conflict has erupted at the local as well as at the national and international levels. Hence, this is as good a point as any to broach the subject of economic trade-offs.

Even with intensive research and some notable technological innovations, air pollution controls greatly increase the cost of doing business. The most directly impacted are the fossil fuel power stations and the oil refineries that must install expensive scrubbers and emission controllers and must also count on operating at lower day-to-day efficiencies. These costs are passed on in short order to energy consumers, other industries, and the public in the form of more expensive gasoline and electricity. Additionally, most industries and businesses are directly affected by the banning of a wide variety of chemical products, from CFCs to paint thinners to cleaning fluids. Not only must they pay for expensive substitutes, but they must also sacrifice performance and efficiency.

In the case of the family car, this means reductions in gas mileage and pickup—aggravations, but not of major proportions. By contrast, the effect on regional economies can be devastating. For a business, the additional cost of implementing pollution controls may spell the difference between profit and loss, success or bankruptcy. For a community it means loss of jobs, growing relief rolls, exodus of the unemployed, declining real estate values, bank failures, etc.—a domino effect that can spell economic catastrophe. On a national level, impaired industrial efficiency may lead to the loss of competitiveness and economic recession, or worse; at least this is the persistent nightmare.

On the other side of the coin is the only-too-real nightmare of the virtual destruction of the landscape and the health of its inhabitants, when environmental concerns are totally subordinated to maximizing industrial production: villages and towns covered by layers of soot, noxious clouds of chemical efflux a permanent presence, countrysides denuded of trees and shrubs. All that has happened in recent years in East Germany, Romania, in parts of the Soviet Union, and many other parts of the world are similarly threatened.

The battle is joined, therefore, between advocacy communities championing freedom from restraint on one hand and the air pollution Community on the other. In many regions of Japan, Europe, and North America, such as Southern California, workable balances have been struck and reasonable progress is being made. But the balance is delicate and is prone to be upset when either side becomes excessively assertive. In my opinion, unqualified predictions of imminent catastrophes, economic or environmental, have that characteristic. While arousing righteous public fervor, they are often counterproductive and inimical to the public interest.

NINE

The AIDS Epidemic

> *If you were the devil, you couldn't conceive*
> *of a disease that would be more disruptive*
> *and disturbing than this one. It could prove*
> *to be the plague of the millennium.*
>
> —DR. ALVIN FRIEDMAN-KIEN[1]

The AIDS threat burst upon the public consciousness in the early 1980s. The disease was not even recognized or defined by the medical establishment until 1981. By 1984, the human immunodeficiency virus (HIV) had been isolated and shown to be the cause of the combination of symptoms known as acquired immune deficiency syndrome or AIDS. By then there was little doubt that the disease had reached epidemic proportions and that a full-fledged catastrophe was in the making.

Prior to 1980 it was widely believed that infectious diseases had ceased to be a threat to public health, at least in the developed portions of the world. AIDS, which is a contagious disease, affecting men and women of all ages and all racial and social backgrounds, came as a devastating surprise. Invariably fatal, AIDS is most often transmitted by sexual contacts, but is also spread by the use of infected needles of drug abusers and by tainted blood supplies. The World Health Organization forecast

193

that the cumulative number of carriers of the HIV throughout the world will exceed 10 million by 1993 and approach 20 million by the year 2000, implying an enormous burden on social services and health care facilities.

By 1990, several hundred books whose titles included the word AIDS were in print. Many of these were specialized texts dealing with medical and social issues. Many of the others were directed to the public in efforts to minimize the further spread of the disease. One of the most widely read general discussions of the subject was J. I. Slaff's and J. K. Brubaker's[2] *The AIDS Epidemic*, published in 1985. A compendium of essays with the same main title, edited by P. O'Malley,[3] appeared in 1990. The October 1988 issue of *Scientific American* was devoted to AIDS. The ten articles and closing essay of that issue were reprinted in book form as *The Science of AIDS*,[4] in 1989.

The AIDS threat is perhaps unique among the catastrophe theories discussed in this book in having inspired serious books with sharply opposing views. Michael Fumento,[5] whose *The Myth of Heterosexual AIDS* appeared in 1990, criticizes the epidemiological projections that have appeared in the scientific literature and suggests that AIDS is far less of a threat than is generally believed. Another critical note was struck by Bruce Nussbaum[6] in his 1990 book *Good Intentions*, which warned that a dangerous conspiracy by big business and the medical establishment was placing profits ahead of science and corrupting the fight against AIDS.

THEORY

Biology. The human immunodeficiency virus, HIV, is a *retrovirus*. It cannot reproduce or replicate by itself but needs to invade another biological structure and take it over. When it does so, it may remain latent for a long time and then abruptly start making many copies of itself. Until 1980, virologists believed that retroviruses occur only in some animals and not in humans. Then in

1980, John Gallo in the United States and others in France were able to isolate the first human retrovirus HTLV-I (for human T-lymphotropic virus type one). Later, Luc Montagnier isolated a related but different virus, HIV, that was actually responsible for most, if not all, cases of AIDS. In the human body, HIV infects certain white blood cells, the T4 (also known as the "helper") cells, and the macrophage. This weakens the immune system and makes the body susceptible to a number of diseases that a healthy immune system is able to control. Two related but different strains of HIV have been isolated. The more widespread of these is designated as HIV-I, while the more recently discovered variety, called HIV-II, is found primarily in West Africa.

Medical Symptoms. The combination of ailments and diseases that are commonly identified with AIDS are actually the late stages of the results of HIV infections. Robert R. Redfield and Donald S. Burke[7] of the Walter Reed Army Medical Center have developed a classification system to characterize the various stages of the HIV-caused disease. Stage 1 is characterized by the absence of any symptoms other than the detection of the HIV in the blood by standard antibody tests. At least six weeks and as much as a year may elapse from exposure to the HIV to the ability to detect antibodies.

Stage 2, which typically lasts from three to five years, is marked by the appearance in the majority of patients of chronically swollen lymph nodes. During Stages 1 and 2 the virus gradually destroys a majority of the T4 cells in the blood. When Stage 3 is reached, the T4 cell count has dropped from a normal 550 or more to below 400 cells per cubic millimeter, at which point the immune system is recognizably impaired. In Stage 4, which commences after another 18 months or so, the immune system fails to respond to various foreign proteins injected under the skin. Stage 5 is often characterized by the first overt symptoms, fungus infections in the mouth and other mucous membranes. By then, the T4 cell count has usually fallen below 200 and may be as low as 50. Within a year or two after entering Stage 5, so-called "opportunistic" infections break out in various parts of the body, and the patient enters the final stage, Stage 6, which is recognized as AIDS.

Few patients survive two years in Stage 6. At this point, the body is unable to resist infections that are usually present in the healthy body. These may cause severe types of pneumonia, cancer, meningitis, encephalitis, and blindness, among many other diseases. A variety of fungal and bacterial infections are quite common. In addition, familiar, treatable diseases, such as tuberculosis, may prove fatal. The prevailing theory is that most, if not all, people infected with HIV eventually reach Stage 6 and will suffer a premature death.

Origin and Spread. A virus closely related to HIV is very common in certain species of African monkeys. Seventy percent of the green monkeys of central Africa appear to be carriers of the virus, but do not develop AIDS or other symptoms. It is conjectured that the virus first evolved in these monkeys and was then somehow transmitted in the 1960s to humans living in relatively isolated forest communities. Eventually, the virus spread to the more heavily populated regions of Africa and was then transported to the European and American continents. By 1990, it was estimated[8] that there were 300,000 AIDS cases in Africa, 250,000 in America, 50,000 in Europe, and 3,000 in Asia and Oceania. The number of individuals infected by HIV was estimated to number many millions, including an estimated two million in the United States alone.

HIV is a blood-borne virus. It is transmitted primarily by blood-to-blood and by semen-to-blood contacts. The transmission appears to be limited to direct contacts with infected bodily fluids. Most infections can be traced to sexual intercourse, the use of unsterile needles by intravenous drug abusers, and to inadequate medical facilities, including some contaminated blood banks. The virus is also passed from infected mothers to their babies. In developed countries, HIV first appeared in homosexuals, but, more recently, an increasing proportion of patients has turned out to be heterosexual.

Treatment and Prevention. As of the time of this writing no cure for HIV infections has been discovered or is in the offing. Nor was there any expectation of an effective vaccine against HIV infection or against AIDS. Some drugs, notably azidothymidine, AZT for short, appear to delay the onset of AIDS symptoms for some time, and others provide symptomatic relief for some of the AIDS-related illnesses. These medications are of great, if only temporary, benefit to certain patients, but they do not constitute steps toward an early conquest of the disease or of the epidemic. For this reason, major efforts are being directed at the prevention of the spread of HIV and at coping with the increasing burden that the disease is expected to place on the health care system.

The various mechanisms that are significant in the spread of HIV infections and of AIDS are shown in Table 6, along with generally recommended preventative methods. The fight against the further spread of the disease has focused on the two principal modes of transmission: "unsafe" sexual contacts and the sharing of needles by intravenous drug users. Public health agencies throughout the world have sponsored intensive campaigns to educate the public about the danger of AIDS and the need for taking adequate precautions. The use of condoms made of synthetic materials is strongly urged for those unwilling or unable to change their sexual habits, and the use of clean injection implements is advised to those who are addicted to intravenous drug use. In some localities and communities, condoms and needles are freely distributed to all who request them.

DATA

The data that are the basis for models and prediction fall into two major categories: (1) Biological and clinical studies of HIV

TABLE 6. Prevention of AIDS Transmission
(from C.B. Wolfsy, in Sande and Volberding,[9] by permission from W. B. Saunders Company)

Transmission route	Transmission mechanism	Fluid or tissue implicated	Prevention
Sex	Male to male, heterosexual, woman to woman	Semen, vaginal or cervical secretions	Careful partner selection; HIV discussion with potential partner; abstinence; latex condoms used properly; water-based lubricants; nonoxynol-9
Intravenous drug use	Needle sharing	Blood	Not starting drug use; getting drug treatment; not sharing needles; cleaning injection implements; cleaning skin; avoiding "shooting galleries"
Blood and blood products	Transfusion, hemophilia, treatment	Blood clotting factors	Voluntary donor exclusion; HIV testing; heat treatment
	Needlestick exposures	Blood	Avoiding recapping needles; using puncture-proof containers; following CDC guidelines
Maternal (seropositive mother)	Transplacental, birth	?Blood, fetal tissue, amniotic fluid	Avoiding conception if HIV positive; terminating of pregnancy; anticipating care of infected child
		Breast milk	Avoiding breast-feeding (in United States)
Sperm donation		Semen	Voluntary donor exclusion; reputable donor center; donor HIV test at 0 and 6 months; freezing sperm
Organ donation		Various organs	HIV testing before donor receives any transfusion; risk benefit analysis

and of AIDS, and (2) epidemiological material to track the number of cases throughout the world and to identify the sectors of the population that appear to be the most susceptible.

Enormous quantities of biological and medical data have been gathered in efforts to identify the cause or causes of AIDS, to establish the manner in which HIV invades the body and destroys the immune system, and to explore the efficacy of various remedial and preventative measures. These evidence steady progress in the understanding of the disease and related medical problems and in providing short-term relief from some of the symptoms. However, as already indicated, neither a vaccine nor a cure has yet been discovered. Data from biological and clinical studies are therefore not useful for the prediction of the eventual extent of the AIDS epidemic or of the impact that it will have on society at large. For this reason the present discussion emphasizes epidemiological data relating to the nature and magnitude of the epidemic.

As noted earlier, the AIDS epidemic started in Africa, which probably still leads all others in the number of HIV infections and cases of AIDS. Unfortunately, health services in general and the facilities to report on diseases in particular are very poor in many parts of Africa. The World Health Organization has made careful estimates of the incidence of AIDS in different parts of the world. Figure 26 shows the number of AIDS cases that had been reported as of March 1, 1990 on each continent along with estimates of the actual number of cases. In some areas, the estimated numbers are based on serologic studies and limited surveillance. Note that in Africa less than 14% of all cases are reported. As a result, it is difficult to make a meaningful breakdown and analysis.

By contrast, very detailed data are available for the United States and parts of Europe. Table 7 shows the number and distribution of AIDS cases reported in the United States as of February 1990. These are cumulative figures compiled by the Centers for Disease Control. Of particular interest and importance in predicting the course of the epidemic in the future are trends in the number of new cases reported month by month or year by year.

TABLE 7.   AIDS Cases Reported in the United States through January, 1990 (from T. C. Quinn, in Sande and Volberding,[12] after Centers for Disease Control: HIV/AIDS Surveillance Report, February 1990, by permission from W. B. Saunders Company)

Transmission category	Males No.	Males %	Females No.	Females %	Total No.	Total %
Adults/Adolescents						
Homosexual/bisexual	72,153	66			72,153	60
Intravenous drug user (IVDU)	19,489	18	5,711	52	25,200	21
Homosexual male/ intravenous drug user	8,326	8			8,326	7
Hemophilia/coagulation disorder	1,071	1	28	0	1,099	1
Heterosexual contact	2,399	2	3,454	31	5,853	5
Transfusion recipient	1,814	2	1,109	10	2,923	2
Other/undetermined	3,286	3	750	7	4,036	3
Subtotal	108,538	100	11,052	100	119,590	100
Children						
Parent with/at risk for AIDS	846	76	829	89	1,675	82
Hemophilia/coagulation disorder	104	9	4	0	108	5
Transfusion recipient	141	13	76	8	217	11
Undetermined	27	2	28	3	55	3
Subtotal	1,118	100	937	100	2,055	100
TOTAL	109,656	100	11,989	100	121,645	100

THE CAUSAL CHAIN

The prediction of the AIDS catastrophe entails the projection of the number of people that are going to be infected with HIV and those that will show the symptoms of AIDS in future years. There are reports of cases of HIV and AIDS in virtually all countries on the globe. But how reliable are they? With reference to the comprehensive model shown in Figure 27, the estimate of people who are actually infected at present is represented by Model A. The output of Model A serves as one of the numerous inputs to Model E, the model that actually predicts the number, and perhaps also the geographic and demographic, distribution of AIDS and HIV cases in the future. Other factors that affect this prediction include the outputs of Models B, C, and D, and probably a number of others. These take into account the effect of changing social patterns and sex habits, the effect of education, the various preventative measures, and the possibility that an effective vaccine may be invented.

In terms of the Spectrum of Models, each of the component models of Figure 27 is governed by biological and social science considerations and, therefore, lies in the dark gray area of the spectrum. The chaining of these submodels results in a comprehensive model with very low predictive validity.

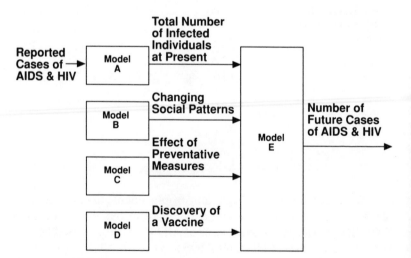

FIGURE 27. Comprehensive model of the AIDS epidemic.

PREDICTIONS

The modeling of the incidence of HIV infections and of AIDS is fraught with uncertainty. Most models are based on the premise that no cures or vaccines for AIDS will become generally available in the foreseeable future, probably a safe if conservative assumption. The forecasting of the course of the epidemic then entails the projection of past trends into the future and the estimation of the possible future beneficial effects of preventative measures on the rate of HIV infections.

The accuracy of estimates of the future incidence of a disease is determined by a number of factors. "Bias" in a model results from attempts to project, on one portion of a population, experience gained from another portion. For example, observations made in New York may not be applicable in San Francisco, and it is up to the modeler to decide to what extent to base predictions made in San Francisco on data collected elsewhere.

Similar judgment is required in "splitting" and in "lumping" data. For example, in Table 7 the number of AIDS cases in the United States has been divided (or split) into twenty-two categories, eleven for males and eleven for females. On the other hand, the data have been lumped geographically; case reports from all parts of the United States have been combined. Each of the categories may have distinctly different dynamics in the future, some will shrink as preventative methods become more effective and others grow as the disease spreads into new sectors of the population.

The organization of Table 7 reflects assumptions made by the modeler as to what is significant and what is not. Some of the categories are very small, the smallest containing only four cases. The percentages associated with these categories are likely to fluctuate substantially from reporting period to reporting period. Other categories are as large as 72,153. Each of these is perhaps better described by a number of subcategories based on considerations such as geography, economics, ethnicity, etc.

According to Victor de Gruttola and William I. Bennet,[13] the modeling of only those aspects of the transmission of AIDS that relate to sexual activities requires a knowledge of the following factors:

- the rate at which the virus is transmitted from one person to another, according to sex of partner, along with each of the potential routes of contagion
- how this rate changes over the course of infection
- a measure of how transmission is influenced by the presence of cofactors, such as other diseases and use of drugs
- the rate at which susceptible partners (sexual, drug using) are acquired
- the frequency and type of sexual contact between such partners
- the average duration of the relationship during which the relevant activities occur
- the extent to which the relationships and activities occur within relatively closed networks, as opposed to extending throughout the population
- the elapsed time between becoming infected and becoming ill

In practice, few if any of these factors are known with sufficient accuracy. Detailed models of the epidemic, therefore, cannot be constructed.

Because of the difficulties inherent in epidemiological modeling, only general predictions have been published, even for the United States where the reporting of new cases is relatively reliable and complete. Worldwide predictions are even more general and tentative. Jonathan M. Mann and his colleagues,[14] writing in 1988, stated, "From our current knowledge of the disease, we estimate that over 250,000 cases of AIDS have already occurred, that between five and 10 million people worldwide are infected with the AIDS virus and that within the next five years about one million new AIDS cases can be expected. In short, the global situation will get much worse before it can be brought under control."

Separate projections and predictions deal with the medical costs of the epidemic. The U.S. Department of Health and Human Services predicted in mid-1991 that the cost of treating HIV infections in the United States would be $5.8 billion in 1991 and grow to $7.2 billion, $8.7 billion, and $10.4 billion in 1992, 1993, and 1994, respectively. About 25% of these annual amounts would be used to treat patients with HIV, but without the symptoms of AIDS. Daniel M. Fox and Emily H. Thomas[15] present information that indicates that initial predictions of the cost of medical treatment forAIDS were substantially inflated. Nonetheless, AIDS promises to place a crushing burden on all components of the United States health care system.

WHAT MUST BE DONE

The crisis has already arrived. And things will get a lot worse. The measures to be taken fall into three categories.

Research. Large-scale research projects have been under way in the United States, Europe, and Japan for a number of years, and these should be continued at greatly increased levels. Some of these entail basic research to gain better understanding of the molecular and cellular biology aspects of the HIV and related infections. These projects have already borne impressive fruit.

A second major research direction is the development of an effective vaccine against HIV. This is certainly the best hope for the eventual conquest of the disease. There are very effective vaccines against other viral diseases including polio, smallpox, yellow fever, and mumps; so the availability of a vaccine could eventually lead to a substantial decline, if not the complete elimination, of AIDS. Despite intensive efforts, progress in this direction has been very slow, and former Surgeon General C. Everett Koop has warned that no satisfactory vaccine can be expected before the end of the century. Thomas J. Matthews and Dani P. Bolognesi[16] describe three obstacles that must be overcome in order to develop an HIV vaccine: ". . . the devious nature of the virus itself, which can 'hide' in cells, change the composition of its coat and install its own genes within the genes of its host; the lack of a good animal model for the disease, which slows the investigation of vaccine strategies to combat these ploys; and the difficulties expected with clinical trials, which face scientific uncertainty, ethical concerns and possibly a shortage of volunteers."

In the meantime, extensive research is in progress to find improved means to give symptomatic relief to AIDS patients and to slow the course of the disease.

Prevention. Extensive efforts have been conducted for several years to educate the public in general and members of "risk groups" in particular about the danger of AIDS and how to avoid HIV infection. These include special programs in schools as well as concerted attempts to reach the adult public. Local programs have been developed to distribute free condoms and free needles for intravenous drug users. Additionally, there are numerous endeavors to make facilities for testing for HIV infections widely available at low cost to alert potentially contagious individuals and encourage them to refrain from high-risk activities. All of these measures have been hampered to a greater or lesser extent by local opponents who fear that these methods condone promiscuous sex and drug use. Nonetheless, these and similar efforts need to be greatly intensified to slow the spread of this disease.

A PERSONAL VIEW

This was the most difficult chapter for me to write. Seven years ago, a member of my family, whom I loved very much, succumbed to AIDS. One of my students died of the disease years after receiving a transfusion of tainted blood. So I know firsthand about the terrible ravages of that disease and the pain that it inflicts on families and the community. For this reason, I usually respond generously to personal appeals for funds to help fight AIDS.

Among the eight "imminent catastrophes" discussed in this book, the AIDS epidemic is widely perceived as extracting the biggest toll at present. Certainly, there are many deaths that are attributable in one way or another to overpopulation in many parts of the world. Air pollution and radiation are probably also directly or indirectly responsible for many deaths and serious illnesses. But these come to our attention only on rare occasions, and then only as parts of much larger problems. By contrast, we are brought face to face with the impact of AIDS almost daily as the lives of more and more prominent people are cut short by the disease. Many people have become alarmed by its gradual spread from so-called "high-risk" groups to the general population, though many people felt compassion for the hundreds of thousands of people making up that high-risk group. When we read about the havoc that it is still wreaking in Africa, we surmise that a new bubonic plague is upon us. Hence, everyone feels personally threatened. The AIDS catastrophe has arrived. The other catastrophes appear still to be waiting in the wings.

So at this point in time, the AIDS calamity tugs on our heartstrings to a degree matched by very few other causes. And the public response has been outstanding. Millions and millions of dollars have been contributed by individuals and by organizations. At the same time, extremely strong political pressure has been brought to bear on the medical establishment and on the government to mount crash programs to cope with the problem. The campaigns mounted by the AIDS Community have become a very powerful force indeed; to some a steamroller that crushes everything in its way.

For my part, I am happy to see that medical researchers are going all out to discover an AIDS vaccine, and I am also in favor of shifting preventative measures into high gear. All money raised from private sources for this cause is money well spent. I am uneasy, however, about the magnitude of the public funds that are

TEN

Overpopulation
The Population Explosion

Arresting global population growth should be second in importance only to avoiding nuclear war on humanity's agenda.

Whatever your cause, it's a lost cause without population control.

—PAUL R. AND ANNE H. EHRLICH[1]

Among the eight imminent catastrophes reviewed in this part of the book, the threat of overpopulation has seniority. Almost two hundred years ago, Robert Malthus forcefully asserted that the population of the world was growing far more rapidly than available food resources. Because of this, he forecast a catastrophic disintegration of living standards. Since that time, a legion of demographers, economists, and political scientists have championed Malthusian doctrines and prescriptions as how best to avoid, or at least to postpone, the predicted catastrophe. To buttress this cause there have also been myriad movements and organizations to promote family planning and birth control.

In modern day America, Paul R. Ehrlich has become at once the elder statesman and the most influential of the scientists currently advocating measures to control the growth of population in all parts of the globe. His best-selling *The Population Bomb*,[2] first published in 1968 and revised in 1971 and 1978, gives an eloquent warning of the dangers posed by the unchecked growth of the population and a call to all people to militate for effective measures to reduce the birth rate. His 1990 best-seller, *The Population Explosion*,[3] co-authored with his wife, Anne H. Ehrlich, reiterates and updates this message and views the problem in the current political context. The book *Extinction*,[4] by the same authors, which appeared in 1981, has been described as *The Silent Spring* of the 1980s in dramatizing the disastrous impact of modern technology and the growth of human populations on the flora and the fauna of the world. In addition to a long series of books, Professor Paul Ehrlich has published several hundred scientific papers on ecology, evolution, and behavior.

The overpopulation threat is unique among the "imminent catastrophes" in that most measures proposed to mitigate it have aroused strong and persistent political opposition, not only in the Third World, but also in the United States and most European countries. As a result, recent progress has been spotty and uneven, despite the fact that the population problem is generally recognized as being at the root of most of the other "imminent catastrophes."

THEORY

The Legacy of Malthus. Robert Malthus, an ordained Anglican clergyman, achieved great renown in the first half of the nineteenth century through a series of essays dealing with the fundamental principles underlying demographics and political economics.[5] Prior to Malthus, it was generally believed that a large and rapidly growing population is a sign of economic prosperity and of national power. The larger the population, the larger the army that can be sent into battle, and the better neighbors can be dominated and colonies acquired. For these reasons, most governments did their best to encourage a high birthrate in their domains. Malthus attacked that point of view.

According to Malthus, populations grow at a geometric rate while food supplies grow at an arithmetic rate. In other words, the increase of population achieved during any year is proportional to the size of the population at the beginning of the year. If the birth and death rates remain constant, the population increase each year will be greater than the population increase for the preceding year. The population, in absolute terms, will therefore grow more and more rapidly as time goes on. In contrast, food supplies increase by the same absolute amount each year; this is similar to the difference between compound and simple interest. From this, Malthus concluded that inevitably the requirements of the growing population would outstrip available food supplies, resulting in famines, diseases, wars, and other social miseries. Opposed to most forms of birth control on moral grounds, Malthus painted a dismal, catastrophic picture of the future.

Over the years, some of Malthus's premises have turned out to be incorrect. The population of the world has, for various reasons, grown more slowly than Malthus predicted. At the same time, advances in agriculture and technology have outpaced the increase of population in many parts of the world. The effect has been to postpone the predicted catastrophe, but not to eliminate the threat. Malthus's theories have also had a profound influence outside the immediate area of demography. His work suggested to Charles Darwin that a process of natural selection leads to the "survival of the fittest," a concept which became a cornerstone of the theory of evolution. Malthus's work also inspired leading political scientists and economists from John Stuart Mill to John Maynard Keynes. As a result, Malthus is still widely regarded as having formulated one of the "great theories of mankind."

The Neo-Malthusians. Even before the advent of the Malthusian doctrine, it was widely recognized that when the population becomes excessively large, consequent famines, epidemics, and wars intervene to limit further growth, albeit at an unacceptable social cost. Advances in agriculture, in medicine, and in technology have worked to postpone the onset of many of these miseries, but they provide no magic solutions. Sooner or later we must still come to grips with potential catastrophes in each of the following areas.

1. *Food*: As time goes on, it is very likely that it will be food shortages that will place a final cap on population growth. Periodic famines have always been a fact of life in many parts of the world. Modern agriculture has greatly increased the amount of food that can, under the right conditions, be produced from an acre of farm land. The so-called "Green Revolution," which introduced new hybrid plants through "genetic engineering," as well as fertilizers and irrigation technologies, had particularly dramatic beneficial effects in Asia in the decades following World War II. In retrospect, however, many of these "advances" extracted a long-term price. More intensive agriculture required the expansion of cultivated land areas and more water. This resulted in deforestation, erosion, mineral depletion of soils, and declining water tables. The Green Revolution bought time, but it did not provide a cure. Even at present population levels, many of the largest countries in the world are often on the brink of starvation. Only the United States, Australia, New Zealand, Argentina, Thailand, and some European countries regularly produce enough food to make substantial exports. And a bad year for them, say due to a drought, means reduced food exports and, therefore, spells trouble all over the world. The famines during the past decade in the Sahel region of Africa are harbingers of things to come.

2. *Environment*: Many of the well-documented insults to the global environment can be directly attributed to overpopulation. The emission of greenhouse gases and the consequent global warming is likely going to increase with catastrophic effects unless population growth is brought under control. To be sure, at present, the developed countries are primarily responsible for this phenomenon, but as the developing countries reach for a higher standard of living, they can be expected to contribute more greenhouse gases

as well as many of the other major pollutants—CFCs, sulfur dioxide, oxides of nitrogen, and perhaps even radioactive particles. Over twenty years ago, Paul Ehrlich[6] introduced a widely quoted equation to determine the environmental impact of increasing populations and rising standards of living:

$$\text{Impact} = \text{Population} \times \text{Affluence} \times \text{Technology}$$

or I = PAT, for short. The first term on the right-hand side of the equation is the total number of people, the second term is a measure of the average per capita consumption of resources, and the third term expresses the disruptiveness of the technologies that generate these resources. At present, the developed countries in Europe and North America have a very high A and a very high T, and there are many proposals directed to reducing these figures. But from a global point of view, a modest increase in the standard of living elsewhere, particularly in China, would certainly overshadow any gains made in this way.

3. *Public Health*: Increasing populations have disastrous effects on the health of people all over the world. Chemical pollution by the fertilizers used to squeeze ever larger crops out of the earth, as well as air pollutants, have been linked to a wide variety of miseries including many forms of cancer and birth defects. Over the centuries epidemics have swept through many overpopulated regions, and some of them have contributed to the decline and the fall of entire civilizations. Today's overpopulation, particularly in Africa, has provided breeding grounds for new and deadly diseases.

4. *Ecology*: Increasing population densities and the resulting environmental pollution exert pressure on plants and animals ranging from microorganisms to elephants.[7] Every year hundreds of species become extinct. This process is greatly exacerbated by the recent destruction of tropical rain forests and by desertification. All of the species now in existence form a global gene pool, and the extinction of any species constitutes an irreversible loss. Many of the species now rapidly becoming extinct are of great benefit or of

potential benefit to mankind. For example, there are some species of wheat that are exceptionally hardy, able to resist various agricultural pests. But these species are disappearing one by one, as farmers turn to other species that provide richer harvests per acre. In the future, as new blights and insect infestations appear, we may well wish for some of the hardy strains and greatly regret that the demands of exploding populations were given priority over conservation.

Neo-Malthus demographers, while echoing Malthus's concerns about the evil consequences of overpopulation, have focused on practical means to limit the growth of population. Here they have had to confront a host of moral and religious issues. In theory, the population can be controlled by limiting the birth rate on one hand and by limiting the average life span on the other. Over the years, attempts to implement the latter have included the systematic exposure and killing of infants, still a population-control practice in some remote parts of the world, as well as the slaughter of "defective" or unwanted children and adults, culminating in the holocausts and genocides of our time. These measures are, of course, rejected out of hand by virtually all contemporary scientists.

At the same time, it must be recognized that many of the modern advances in medicine, public health, and public services are directed to increasing the life span of all individuals. The very measures that we applaud as improving the quality of our lives are also exacerbating the overpopulation problem. In a sense, the arrival of any of the other seven "imminent catastrophes" would improve the overpopulation picture. On moral, ethical, and religious grounds, most scientists choose to spurn any artificial steps to reduce or to limit anyone's life expectancy and fully support progress in medicine and public health. The control of overpopulation is therefore generally viewed as being synonymous with birth control. The question is: How?

The Anti-Neo-Malthusians. The beliefs and recommendations of the neo-Malthusians, as described above, have come under severe criticism not only on religious and political grounds but also from the theoretical point of view. A number of prominent economists have argued that population growth is not necessarily detrimental to the general welfare and will, in fact, lead to greater prosperity in the future, provided the economy is properly managed. Julian Simon's and Herman Kahn's[8] *The Resourceful Earth* is one example of the writings of the opponents of the overpopulation catastrophe theory. They are, of course, not members of the Community concerned with overpopulation, and therefore fall outside the scope of this book.

DATA

Demographers love numbers, and the technical literature overflows with demographic data—the population in different parts of the world throughout history; that is, how the people lived and how they died. Tons of data are gathered and compiled by various United Nations agencies, by the World Bank, by most national governments, and by many scientific organizations. For the past sixty years, the Population Reference Bureau (PRB), a nonprofit corporation based in Washington, D.C., has been doing yeoman work in digesting this information and making it accessible to the general public. The data presented below are taken from various PRB publications.

The set of numbers that jumps out at you from most world population reports and the one that regularly gets the most attention is the tabulation of the total population of the world as it has evolved since the beginnings of human history. This evolution is portrayed dramatically in Figure 28, which depicts vividly the meaning of the expression "population explosion," and suggests that a catastrophe is indeed imminent unless present trends are promptly reversed.

Most modern demographers are quick to point out that Figure 28 conceals nearly as much as it reveals. To make meaningful inferences from the raw populations, much more detail is needed: the geographic distribution of people all over the world,

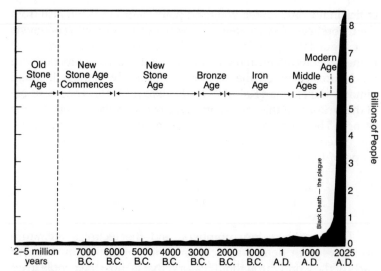

FIGURE 28. World population growth through history (provided by Population Reference Bureau, Inc.,[9] by permission).

the distribution of their ages, the ratio of men to women, their economic status, and much, much more. Only with this additional knowledge in hand can one make reasonable projections into the future.

Thomas Merrick,[10] president of the PRB, in his 1989 monograph, "World Population in Transition," identifies two periods of major population growth: 1750 to 1950 and 1950 to the present. Until 1750, the birth rate was high everywhere, but so was the death rate, due to generally poor health and nutrition, famines, epidemics, and wars—perhaps roughly in line with Malthus's theory. Even so, the world population reached about 760 million in 1750.

After 1750, most relatively developed countries sooner or later began to experience a so-called *demographic transition*, at which time the death rate dropped sharply as a result of improved medical, social, and economic conditions. Meanwhile, the birth rate

rate declined less rapidly as living standards improved. So during that period the population increased substantially. By the end of this transition, the population in most developed countries had leveled off or was growing only very slowly, as the declining birth rate caught up with the declining death rate. In the meantime the world population grew to 2.5 billion by 1950, due in part to the growth in the developed nations. Between 1950 and the mid-1980s the world population doubled to five billion, largely as a result of the surge in the growth rates in the less developed and under-developed countries of Asia, Africa, and Latin America. This is summarized in Tables 8 and 9.

Table 10 is a snapshot of the situation as of 1990. The following are definitions of some of the terms in that table, as provided by the PRB:[11]

- *Rate of natural increase*: birth rate minus death rate
- *Infant mortality rate*: annual deaths of infants under one year of age per 1000 live births
- *Total fertility rate*: average number of children a woman will bear in her lifetime under the present mortality conditions
- *Life expectancy*: average number of years a person can expect to live under the present mortality conditions

The distribution of the ages of the residents of a specific country is an important indication of the extent to which its demographic transition has been completed. The greater the pro-portion of old people, the smaller is the gap between the birth rate and the death rate, and the slower is the population growth. Figure 29 shows the so-called *population pyramids* for the estimated age distributions in 1985 for three countries in different stages of the demographic transition. Mexico with a large majority of young people is experiencing rapid population growth; the United States is in the so-called "constrictive" stage in which the growth is slow and a smaller percentage of the people are young; Sweden has arrived at the "stationary" or near zero-growth stage with roughly equal numbers of people in all age ranges.

The recent news has not been all bad. In 1991, there were some favorable signals. The Demographic and Health Surveys funded

TABLE 8.   Population Growth in Five Major Regions
Before and After 1950 (from T. W. Merrick,[12]
by permission from Population Reference Bureau, Inc.)

	1750–1950		1950–1985	
	Number in millions	Percent	Number in millions	Percent
World total	1,756	231.1	2,321	92.2
Less developed regions	1,112	195.4	1,976	117.5
Africa	124	124.0	331	147.8
Asia (minus Japan)	837	184.0	1,405	108.7
Latin America	151	1,078.6	240	145.5
More developed regions	644	337.2	346	41.4
Europe, USSR, Japan, Oceania	480	254.0	248	37.1
U.S., Canada	164	8,200.0	98	59.0

by the U.S. Agency for International Development indicated that
fertility rates in the developing world have declined by about one-
third since 1960—to 4.2 births per woman from 6.1 births per
woman. The fertility decline in the twenty-eight countries covered
by the survey was very uneven. Some sample fertility rates in
births per woman are: China and Korea, 2.3; Southeast Asia and
Latin America, 4.0 to 5.0; South Asia, West Asia, and North
Africa, 5.0 to 6.0; Sub-Sahara Africa, 6.4. This is encouraging, but
in order to stabilize the world population, the fertility rate would
have to decline to about 2.0, and only China and Korea are
approaching that figure.

THE CAUSAL CHAIN

The forecasting of the world population primarily entails the
prediction of future birth rates and death rates. As suggested in

TABLE 9. World Population Distribution in 1750, 1900, 1950, and 1985 (from T. W. Merrick,[12] by permission from Population Reference Bureau, Inc.)

	Total population (numbers in millions)							
	1750		1900		1950		1985	
Region	Number	Percent	Number	Percent	Number	Percent	Number	Percent
World total	760	100.0	1,630	100.0	2,516	100.0	4,837	100.0
Less developed regions	569	74.9	1,070	65.6	1,681	66.8	3,657	75.6
Africa	100	13.2	133	8.2	224	8.9	555	11.5
Asia (minus Japan)	455	59.9	867	53.2	1,292	51.4	2,697	55.8
Latin America	14	1.8	70	4.3	165	6.6	405	8.4
More developed regions	191	25.1	560	34.4	835	33.2	1,181	24.4
Europe, USSR, Japan, Oceania	189	24.9	478	29.3	669	26.6	917	19.0
U.S., Canada	2	0.3	82	5.0	166	6.6	264	5.5

TABLE 10. The World Population Statistics in 1990
(from Population Reference Bureau, Inc.,[13] by permission)

	World	More developed	Less developed	Less developed (excl. China)
Population estimate mid-1990 (millions)	5,321	1,214	4,107	2,987
Birth rate (per 1,000 pop.)	27	15	31	35
Death rate (per 1,000 pop.)	10	9	10	11
Natural increase (annual, %)	1.8	0.5	2.1	2.4
Population "doubling time" in years (at current rate)	39	128	33	29
Infant mortality rate	73	16	81	91
Total fertility rate	3.5	2.0	4.0	4.6
% Population under age 15/over 65	33/6	22/12	36/4	40/4
Life expectancy at birth (years)	64	74	61	59
Urban population (%)	41	73	32	36

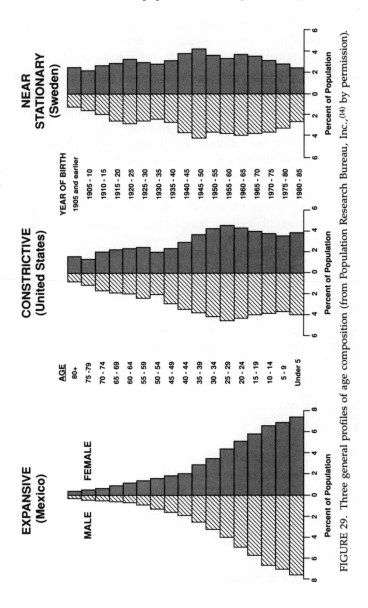

FIGURE 29. Three general profiles of age composition (from Population Research Bureau, Inc.,[14] by permission).

Figure 30, both of these variables are strongly affected by a variety of demographic, scientific, social, and economic factors.

Demographers[11] find all of the following considerations to be among those vital to the projection of population trends:

1. Age composition: the percentage of people falling into each of about twenty age brackets, with separate records for males and females, as in Figure 29.
2. Fertility: the number of children a woman between the ages of fifteen and forty-nine can be expected to bear, allowing for some childhood mortality. A fertility rate of 2.1 children per couple is considered the "replacement level" where the number of births equals just about the number of deaths; above 2.1 the population grows, below 2.1 it shrinks.

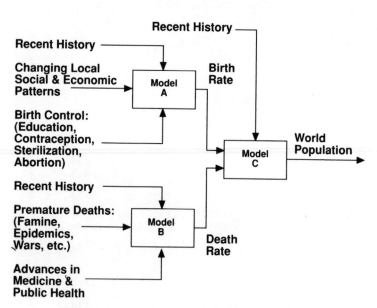

FIGURE 30. Comprehensive model of world population.

3. Fertility rate: the number of live births per thousand women aged fifteen to forty-nine in any given year.
4. Infant mortality rate: number of deaths of infants per thousand live births in any given year.
5. Life expectancy: the average number of years a person would live if current mortality trends were to continue unchanged.
6. Maternal mortality rate: number of deaths of mothers during pregnancy and childbirth per 100,000 live births.
7. Migration: the number of people crossing boundaries to establish a new permanent residence.
8. Morbidity: the frequency of diseases and illnesses.
9. Nuptiality: the frequency, characteristics, and dissolution of marriages.
10. Population density: the number of people per square mile.
11. Urbanization: growth in the proportion of people living in urban areas.

All of these and many more must be known not only for the world as a whole, but for every region and country and usually for the different ethnic, social, and economic groups within each country. And many of these depend in turn on such factors as:

1. Resource management all over the globe, particularly of food and energy resources
2. Advances in medicine as they affect the life expectancy
3. Advances in science and technology as they affect the standard of living
4. Government policies relating to birth control

There is also the problem of *population momentum*. When the population growth of a country or region falls to the replacement level of 2.1 children per couple, the number of births and deaths eventually balance out; but not right away. As described by Merrick,[15] past demographic patterns such as fertility, mortality, and migration play an important role. He compares two populations that both crossed the replacement level threshold of 2.1 around 1985. The population of China will grow by over 50% before reaching equilibrium, while that of Yugoslavia will grow by less than 30% in the future, principally because in 1985 China had a much larger percentage of people under 15 years of age than did Yugoslavia.

Looking again at the comprehensive model of Figure 30, we note first the immense number of variables in making global predictions—the "curse of dimensionality" in spades. Additionally, we are continuously moving into new and uncharted terrains as far as the world population is concerned. Clearly, we are also far "out of bounds." In the terms of the discussion of Chapter 4, it is impossible to include all the variables (dimensions) that may be significant in the model. Moreover, the population is predicted to assume magnitudes never before reached (and, therefore, beyond previously established bounds on validity), which means that the model cannot be validated. It follows that the population model is certainly in the very dark gray region of the Spectrum of Models. Still, in the short run, some population models have proven to be surprisingly accurate. For example, predictions made around 1958 were right on the mark as far as the population growth during the past three decades is concerned. Of course, now we are not just projecting for the next thirty years, but the next one hundred years.

PREDICTIONS

Recognizing the thinness of the ice on which they are skating, most demographers carefully refrain from making "predictions" that imply quantitative forecasts, and prefer to talk instead of "projections"—another word for "scenarios." Figure 31 shows five different projections made by the United Nations. The fertility assumption, the year when the replacement level of 2.1 will be reached, is different for each projection, yet all five "predict" zero population growth at some point in the future. The *when* de-

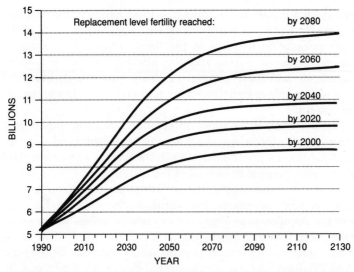

FIGURE 31. Ultimate world population under five different fertility assumptions (from Population Reference Bureau, Inc.,[16] by permission).

pends on the many assumptions that affect future birth and death rates. At best we can look forward to a world population of about 8.75 billion, at worst close to 14 billion. Take your pick.

PRB demographer Leon Bouvier[17] went out on a limb in 1984 and ventured a prediction of the world population in 2034 and how the population would then be distributed over the globe. He had to assume many things—no nuclear wars, continued exponential growth in science and technology, sufficient food and energy resources, worldwide use of contraceptive devices, and a well-regulated world economy. Many would strongly argue with these assumptions, but they led Bouvier to opt for one of the more optimistic of the United Nations projections. His prediction for the world population on July 1, 2034 is a shade over 8 billion. This is twice the population in 1976 and represents an average growth rate over that period of 1.2% per year (compared with around 1.8% at present). He foresees some difficulties in adjusting to this population growth, but no out-and-out catastrophes. About 15% of the world's population will live in the "least developed countries," including some in Africa and the Middle East as well as in Bangladesh, and these people will truly have rough lives. All others will be able to keep their nose above the water to a greater or lesser extent.

But Bouvier's is the most optimistic view. More pessimistic demographers, including the Ehrlichs, are convinced that we cannot assume a substantial decline in the average growth rate and that we must prepare to face all of the catastrophic effects of this growth on the food supply, the environment, the public health, and the ecology we enumerated earlier in this chapter. They see the world population as being out of control, unless major catastrophes keep it from growing indefinitely. If current fertility levels are not reduced, the world population in the year 2100 would exceed 50 billion—a catastrophe of mind-boggling proportions. So the overpopulation Community calls for a crash program now to head off the coming tragedy.

WHAT MUST BE DONE

The principal goal of the Community concerned with overpopulation is to reduce the birth rate to the point that the population of the globe stops growing and perhaps, eventually, decreases. It is generally recognized that this requires the adoption of a combination of steps and measures. Because some of these run counter to religious and cultural traditions in different cultures, they must be carefully tuned to local circumstances. In theory, all of the following can be effective:

1. *Education*: The starting point for all birth control programs is the education of the general population and its political leaders to inform and convince them of the negative aspects of population growth and of the importance of birth control.

2. *Family planning*: This is a specific form of education directed at married couples, and under certain conditions to unmarried individuals, to help them limit the number of their offspring to two or less per couple.

3. *Contraception*: There is a multitude of devices, drugs, and techniques to prevent fertilization as a result of sexual intercourse. These include well-known mechanical artifices such as the condom, the diaphragm, and the intrauterine device (IUD). Birth control pills have been available in the United States since 1960, and many new medications have been developed since then, notably the French drug RU 486. There are also over-the-counter products such as spermicidal jellies and sponges. Each of these strikes a different balance between effectiveness in preventing conception and safety. Over the years, many once-promising contraceptives have been found to have dangerous, even fatal, side effects and have been withdrawn. Still, by and large, most forms of contraception are considered safe and effective.

4. *Sterilization*: Relatively simple surgical procedures are available to make men and women infertile. These operations are generally free of unwanted side effects, but they are often, though by no means always, irreversible.

5. *Abortion*: The termination of a pregnancy by chemical or surgical means is by far the most controversial of available measures to reduce the birth rate. In most areas this approach is rejected on religious and moral counts and not because of medical considerations.

POLITICS

The advocates of birth control have encountered heavy political opposition at every step along the way. The end as well as the means have come under heavy attack, and the arguments are too complex to be treated here in detail.

Representatives of the underdeveloped countries have on

numerous occasions asserted that efforts to limit the world population are part of a plot by the developing countries to prevent the rest of the world from gaining its rightful place in the sun. They believe that more people means more power, and they view current demographic trends as tipping the balance more and more in favor of the have-nots. A variety of Marxist and nationalistic arguments are offered to buttress that position.

As for birth control itself, opposition comes from most of the world's organized religions. Devout Christians, Orthodox Jews, Muslims, and Hindus all tend to see efforts to reduce the birth rate as interfering with the will of God. As a result there has always been persistent and concerted opposition to all forms of birth control. Much to the chagrin of the overpopulation Community, the opponents have succeeded in effectively thwarting major birth control projects in many parts of the world.

A PERSONAL VIEW

Overpopulation is the fountainhead of most of the other catastrophes discussed in this book. If only the world population were to become stable at, say, 50% or 75% of its present level, most environmental and public health problems would become more easy to manage. Possibly, we might then be faced with economic stagnation; but that, I believe, could be handled. On the other hand, if the world population continues to grow at its present rate, a plethora of catastrophes, including those represented by the Four Horsemen of the Apocalypse, will be certain to overtake us sooner or later. Ironically, these catastrophes will serve as feedback mechanisms to limit the population, albeit at a terrible cost in human suffering. For this reason, I believe that overpopulation is *the* most crucial global problem that we face today.

Tragically for the human race, most efforts to control the population run counter to the two most dynamic social movements of the 1990s: religious fundamentalism and strident nationalism. Human life is sacred to most of the world's religions, and in the eyes of the more orthodox among the believers, this definitely precludes all forms of abortion. Many of the religious fundamentalists also regard all artificial means of preventing conception as contrivances designed to thwart the will of God and therefore abhor them. During the past decades, fundamentalism has acquired new and devoted followers numbering in the hundreds of millions, particularly in the Muslim regions of Asia and Africa, but also among the Christian majorities in many European and American countries. And the tide still seems to be running strongly in their direction.

In opposing population controls, religious fundamentalists are joined in many parts of the world by ardent nationalists and champions of ethnic minorities. The leaders of most of the developed countries in Western Europe and North America equate population control with improving standards of living and therefore welcome declining birth rates. In the Third World, however, large and growing populations spell increased military power and greater political influence. Therefore, the underdeveloped and developing countries have, for the most part, been active opponents of birth control. This attitude now appears to be spreading to the more developed regions as well. The recent political turbulence in what was formerly the Soviet Union and in Eastern Europe has given impetus to a host of national and ethnic movements, each striving for greater autonomy for its members, greater security from incursion or attack, and often dominance over its neighbors. The temptation is to seek strength through numbers and therefore to favor growing populations, even at the expense of prosperity and welfare.

There is a unique aspect to the overpopulation problem. In the other seven imminent catastrophes, scientists are key players both in predicting the catastrophe and in prescribing and developing the measures for its mitigation. There is contention, of course, between the science-based Community striving to minimize the impact of the catastrophe and other interests more concerned with the high cost of these measures; but both sides look to scientists to come up with practical solutions. For example, in dealing with the greenhouse effect and with the acid rain problems, scientists are advocating conservation to reduce the emission of pollutants from fossil fuel plants and at the same time devising promising alternative techniques for generating electric power economically and without damaging the environment. If and when such techniques are perfected, they will be welcomed by all sides. The situation is different when it comes to overpopulation. Here scientists have been instrumental in alerting governments and the public to the problems posed by the population explosion. They have been generally successful and effective in projecting or predicting the growth of the world population. But when it comes to doing something about it, the issue is not how economical or effective the various methods for birth control are, but whether they should be used at all.

Scientists have developed many reliable and generally safe contraceptives, and in all probability even more powerful drugs will be introduced in coming years. But these can have an impact on the population explosion only if they are widely accepted and used, particularly in regions in which the birth rate is very high. Unfortunately, they are not extensively used precisely in those countries that could derive the greatest benefit from them. This problem cannot be solved by achieving further scientific breakthroughs, nor by the inspired application of scientific models and the scientific method. The solution is in the hands of the political and spiritual leaders all over the world.

I conclude with great regret that it is unrealistic to expect governments, particularly those in the developing countries, to take effective measures to reduce the birth rate while at the same time trying to cope with the growing religious, racial, and national schisms within their borders. In my view, the world population will increase at a rate well above the 1.2% per year projected by L. Bouvier[17] discussed earlier. Therefore, I believe that it would be unrealistic to project a world population under twelve billion by the year 2034. Planners of long-range strategies designed to cope with the problems relating to the environment, to energy and food resources, and to public health should keep the likelihood of these staggering increases clearly in mind.

ELEVEN

Economic Collapse
Another "Great Depression"

*. . . The Great Depression afflicted the world
in 1929 and lasted more than eight years.
I believe that a disaster of the same, if not
greater, severity is already in the making.
It will occur in 1990 and plague the world
through 1996.*

—R. Batra[1]

*The die is cast. We will crash into the worst
economic times imaginable! The crisis is
inevitable.*

—J. L. King[2]

The inclusion of a predicted economic disaster among the "imminent catastrophes" may come as a surprise to some readers. After all, "it's only money." A moment of reflection, however, should convince us that human suffering, including fatalities, that would be caused by a worldwide depression is surely of the same order of magnitude as the potential miseries inflicted by the other catastrophes. Widespread famines, riots, even revolutions and wars might well result from an international economic collapse.

Moreover, since the recommended actions against the other seven imminent catastrophes, without exception, require vast infusions of money, the struggle against them would surely be handicapped, if not doomed, by a calamitous economic downturn. To be sure, the economic catastrophe differs in a number of significant respects from the others discussed in this part of the book.

First, economists have had a genuine catastrophe in the recent past to serve as a model for a similar or worse calamity in the future: 1929, *the* crash. Such a yardstick is lacking when it comes to the other seven imminent catastrophes. Second, catastrophe prediction is not mainstream economics. The prediction of each of the catastrophes described in the preceding seven chapters has the backing, to a greater or lesser extent, of the pertinent scientific establishment. Climatologists may dispute the precise magnitude and timing of the greenhouse effect, but an overwhelming majority is convinced of the dire consequences of the accumulation of atmospheric greenhouse gases. Most of them see a catastrophe coming. The same applies to the effect of CFCs on the ozone layer, the effect of acid rain on flora and fauna, the dangers of radioactive fallout, the impact of the AIDS epidemic, and the dangers of overpopulation.

In economics it's different. Most economists are very uneasy about the state of the U.S. economy, the burgeoning national debt, the export/import imbalance, inflation, and so forth. All expect business cycles to continue as well as consequent fluctuations in the economic variables we use to characterize financial health and malaise. Many believe that we are in for serious recessions, even a depression. Yet, only a relatively small cohort is convinced that we are in for a shakeout as bad and maybe much worse than the "great depression" of the 1930s.

That group of economists, financial advisers, and commentators forms a tight Community—some would call it a cottage industry. They hold frequent conferences and seminars; they are prominent in congressional hearings and in the media; they publish widely read newsletters; and they author a continuous stream of "doom-and-gloom" books that frequently rise to the top of best-seller lists. Each of these books forecasts the onset of very bad times *always within two or three years of publication*. To date, all have missed the mark, and yet they keep selling. The following titles[1-11] and subtitles of some of the more influential of these books will give you a flavor of what has appeared in the past decade: (1979) *How to Prosper During the Coming Bad Years*; (1980) *Crisis Investing: Opportunities and Profits in the Coming Great Depression*; (1980) *The Coming Currency Collapse and What You Can Do About It*; (1985) *The Warning: The Coming Great Crash in the Stock Market*; (1985, 1987, 1988) *The Great Depression of 1990*; (1987) *Apocalypse 2000: Economic Breakdown and the Suicide of Democracy*; (1988) *How to Profit from the Next Great Depression*; (1988) *What's Next? How to Prepare Yourself for the Crash of '89*; (1988) *Apocalypse on Wall Street*; (1988) *Surviving the Great Depression of 1990*; (1989) *The Economic Time Bomb: How You Can Profit from the Emerging Crisis*.

And that is just a sampling; there are many, many more books. But the above serves well enough to show the persistent and the perennial character of this genre. It also helps to highlight another unique attribute of this form of catastrophe prediction. All the books that we have discussed in the preceding chapters forecast catastrophes and exhort their readers to do their utmost to prevent or at least to mitigate the oncoming calamity. Their premise is that we are all in the same boat, that we will all suffer, and that we must all work together to do what we can.

By contrast, in the economics area, while the books predicting a catastrophe often include a chapter or two of gratuitous advice to the U.S. government—raise taxes, control the supply of money, cut the deficit, etc.—they all assume that their advice will fall on deaf ears, and that it is too late in any case to head off the catastrophe. Instead, their primary motivation is to show their readers, the select few, how to escape the impact of the catastrophe and even how to make a few bucks in the process.

It could be that this attitude only reflects the rugged individualism that is a tenet of most conservative economists. But it could also be a sign of things to come in other areas, a leading indicator. Perhaps we should look forward to popular books entitled *How to*

Prosper in the Greenhouse, or *Stocks to Watch as the Hole in the Sky Grows*, or even *How I Made a Killing in Radioactivity*. I certainly hope not.

THEORIES

Over the years, mainstream economists have developed many techniques for analyzing economic data and for using observations of the past to infer the future. Textbooks on the subject, for example, C. W. J. Granger's[12] *Forecasting in Business and Economics*, describe two general approaches: time series analysis and econometrics. In the former, data, such as the daily closing Dow Jones Industrials Index, are processed in order to spot trends and to predict how the average will move in the future. In the second approach, observations of the past dynamics of a number of different variables, such as interest rates, unemployment, money supply, etc., are entered into the computer to determine how they are related to each other and to use these relationships to predict the future. In both cases, the concept of causality plays a vital role, and both have been reasonably successful in forecasting economic trends in relatively normal times.

The forecasters and tipsters of economic catastrophes reach their conclusions via all kinds of models and methods, some of them quite arcane, even occult. Those who purport to approach the subject scientifically, as do all of the ones listed above, generally employ one or the other of the following rationales:

1. The U.S. economy is like a house of cards. Past fatal errors in managing the economy have put us in an extremely critical position. Our monetary system, our banks, the very underpinnings of our economic system have been hopelessly undermined. And everything is going to come crashing down—within two years.

2. A close look at what has been happening recently in the economy reveals striking similarities with what took place just

prior to the 1929 crash. History is going to repeat itself, and we are sure to head into another "great depression," but probably a much worse one—within two years.

The House of Cards. In gauging the state of the economy at present, the "present" being whenever the catastrophe prediction is published (be it 1979 or 1990 or any years in between), catastrophe theorists usually concentrate on several economic variables that they consider to be keys to understanding what is going on. The three most widely mentioned are:

1. *Federal Debt*: The danger of continual and excessive borrowing by the U.S. government has been described by fiscal conservatives since the days of the New Deal before World War II. To a layman, the fact that the national debt increased from under $200 billion in 1940 to over three trillion dollars in 1990 is mind-boggling. That no catastrophic economic dislocations have as yet resulted from this spending binge runs counter to the Puritan ethic and common sense. Forty years ago, virtually no respectable economist would have predicted the continuing relative prosperity throughout the post-war era, given the borrowing and spending policies of successive administrations. But sooner or later the piper will have to be paid. The Japanese, Arab, and European investors who have supported the U.S. government by buying its bonds will run out of patience, will stop buying more bonds, and will want some of their money back. Soon. And then, look out.

2. *Money Supply*: The economy as a whole is very sensitive to the amount of money in circulation. The Federal Reserve Board controls this variable and therefore has its hand on the throttle. Stimulating the economy by permitting the money supply to grow makes for good times and gives an illusion of prosperity, but only for a while. Then prices go up and inflation wipes out all the "gains" that have been made. Hence, the government is forced to reduce the money supply and that results in recession. We are on a recession–inflation seesaw. During the late 1970s and early 1980s,

the government maintained tight control over the money supply and was able to lick inflation for a time. Then, reduced tax collections and increased defense spending during the Reagan era upset the applecart. The economy has been out of control for some time. Therefore, a crash would appear inevitable—and soon.

3. *Bank Futures*: The U.S. banking system has been skating on thin ice for a long time. Credit has been extended to unstable Third World countries, shady real estate speculators, corporate raiders, corporations on an acquisition and merger binge, and to many other unsavory characters. Many of our banks are already bankrupt or nearing bankruptcy, and many more will fail in the near future. The U.S. government has expended trillions of dollars to bail out failing savings and loan associations, but it will be unable to prevent the collapse of the whole banking system when times get rough. And that will cause everything else to come crashing down—soon.

So the reasoning goes. Most economists, not only the catastrophe theorists, see great danger ahead. But will a calamitous crash actually occur? And, if so, when? Predictions of this kind are "outside the bounds" of most economic models. Such forecasts cannot be made with confidence by extrapolating time series or by using tested econometric principles. Instead, catastrophe theorists look to the past, to parallels with the disaster of 1929.

The 1929 Parallel. Although more than half a century has passed, the trauma of the stock market crash of 1929 and the ensuing decade of worldwide depression continues to haunt economists of all persuasions. No present-day catastrophe theorists fails to make prominent mention of the "great depression," and most forecast a replay of the 1929 scenario. Many hold to the theory that when present-day economic conditions have become sufficiently similar to those that existed in the late 1920s, the great crash is sure to come.

Some forecasters base their entire rationale on the 1929/present-day analogy. For example, Joseph Granville's[6] widely read 1985 book, *The Warning: The Coming Great Crash in the Stock Market*, is devoted entirely to 184 parallels between the years preceding 1929 and the years preceding 1985. As a matter of fact, Joseph Granville had used that kind of reasoning to predict a crash in the summer of 1982, when, actually, August 1982 marked the beginning of a sustained five-year bull market. (At this point the reader may wish to reread the section entitled "The Curse of Dimensionality" in Chapter 4.)

Ravi Batra, an economist with impressive academic credentials, used a similar approach to predict the start of a depression in 1990. His books became best-sellers in the late 1980s. Although more concerned with economic fundamentals than Granville, Batra's theory is also largely based on economic cycles and the 1929 model. In his 1988 book,[10] *Surviving the Great Depression of 1990*, he summarizes the similarities that he observed between the 1920–1927 period and the 1980–1987 years. He notes that as far as a number of important economic indicators are concerned—indicators such as interest rates, the number of bank failures, the stock market, unemployment, and inflation—the U.S. economy behaved very similarly in 1920 and 1980, 1921 and 1981, 1922 and 1982, and so on right up to 1929 and 1989. This led him to predict an economic catastrophe for 1990.

DATA

In the case of the other seven imminent catastrophes, theorists continue to be hampered by the scarcity of reliable data, of sufficiently detailed observations of what happened in the past. This is not the situation in economics. Vast amounts of facts and statistics, mostly published by the federal government, are readily available. The problem is what to make of all this information. In his book, *The Economic Time Bomb*, Harry Browne[11] argues that monetary crises result from persistent inflation due to the incorrect control of the amount of money in circulation. To substantiate his view that a catastrophe is at hand, he presents detailed graphs, covering a two-hundred-year period in the United States, of the rate of inflation and of consumer prices. He also plots the number of bank failures and growth in the money supply over a fifty-year period.

What can we learn from data dealing with previous economic catastrophes or near catastrophes? Charles R. Kindleberger[13] in his *Manias, Panics, and Crashes* analyzes over thirty financial crises that have plagued Europe and America from 1720 to 1987 and observes many common features and characteristics. However, most catastrophe theories are based primarily on parallels with 1929.

THE CAUSAL CHAIN

As already stated, many forecasters of the economic catastrophe rely almost entirely on observed similarities of economic conditions in 1929 and the present. All refer to the 1929 crash in one way or another. But quite a few base their predictions, in part, upon presumed causal relationships.

For example, Ravi Batra[14] considers the following fundamental economic factors to have been the principal causes of our hopeless plight at present.

1. The gargantuan U.S. federal debt as a percentage of the gross national product (GNP);
2. The continuing gigantic federal deficits as a percentage of the GNP;
3. The U.S. foreign trade debt and deficit resulting from continuing unfavorable import/export ratios;
4. The growing consumer and corporate debt;
5. The growing Third World debt;
6. Excessive speculation on the stock and commodity exchanges;
7. The shrinking middle class and rising homelessness and poverty;

8. The growing concentration of wealth, as reflected by the increasing share of wealth held by 1% of U.S. adults and families.

Many of these factors are interrelated in complex ways. Most catastrophe theorists agree that all of the above are undesirable, though some categorize a few of them as "nonproblems." But none offers detailed models that would permit the pinpointing of the imminent catastrophe on the basis of observed economic indicators. The economy is just too complex for that. Hence, no credible comprehensive model can be formulated.

PREDICTIONS

All catastrophe theorists agree that we are heading for a precipice. There is disagreement as to the severity of the crash. Paul Erdman[8] foresees a severe but short recession, with relatively good times a couple of years later. Howard Ruff[3] prophesies total chaos and advocates "survivalism"—store enough food for one year, guns and bullets, etc. The others predict something in between. Almost all predict:

- A crash in the stock market, with stocks and bonds losing at least 50% of their value,
- Inflation, perhaps even hyperinflation,
- Very extensive bank failures,
- Collapse of real estate values,
- A sharp decline in the value of the dollar and a sharp increase in the price of gold,
- A general loss of confidence in the government of the United States and in its institutions.

Some of these things have already happened to some extent; but they are predicted to become much more serious. It should be recognized that while the segment of the economic field discussed in this chapter has been crying wolf for the past decades, others have been consistently optimistic. For example, Herman Kahn's[15] *The Coming Boom*, published in 1982, foresaw only good times ahead. And many other futurologists have published similar analyses since then.

WHAT SHOULD BE DONE

A substantial portion of each of the popular books that predict an economic catastrophe is devoted to suggestions and tips. These are designed to help the reader to weather the coming storm and to emerge in good shape economically. People with money to invest are advised to reduce their stocks and bonds portfolio and to avoid real estate altogether. U.S. government debt instruments such as T-bills, Swiss francs, and gold are favored by many advisers.

Many of the investment programs are frankly recommended as disaster strategies. They are designed to play it safe, even if that means losing out on some good investment opportunities. But that can also be expensive in the long run. Had you followed the advice of the catastrophe crowd in the late 1970s, for example, and converted most of your nest egg into gold, you would have been deliriously happy for a few years in the early 1980s when the price of gold shot up to over $800 an ounce. But it promptly dropped back to below $400. In the meantime, you would have been renting instead of owning your own home—when, in some parts of the country, the value of your house may well have increased four- or fivefold in the past ten years. So, what price insurance?

A PERSONAL VIEW

Scientific models of the national and world economy are well-established tools for policy analysis and for planning. Although they fall into the darkest regions of the spectrum of mathematical models discussed in Chapter 3, they have been immensely useful in providing insights into the interrelation of the diverse economic variables, and they have on occasion proven to be reasonably accurate predictors of future events. But only in times of relative calm.

In this discussion we are not concerned with economic models in general, but only with the use of these models to predict an out-and-out catastrophe. All that I know about models and all that I have observed leads me to conclude that catastrophe predictions based on economic models are almost completely devoid of credibility. Sooner or later a genuine economic catastrophe will descend upon us, and some fortunate author will have predicted it more or less correctly and will therefore be lauded as a true prophet and a genius. But in my view, his success will have nothing to do with his skill or his foresight. It will be strictly a matter of luck.

Some recent books dealing with imminent economic catastrophes have been included in the reference listing for this chapter (see Refs. 4, 5, 7, 9, and 16).

TWELVE

Earthquakes
The Big One Is Coming

*Some day in the not-too-distant future, much
of greater Los Angeles will be destroyed. . . .
The toll in human suffering will be
tremendous.*

—DAVID RITCHIE[1]

The subject of catastrophic earthquakes really falls outside the
scope of this book, as carefully defined earlier. First, earthquakes
are definitely not global in character. They are local in their impact,
each major quake affecting localities within a radius measured in
tens of miles. Moreover, disastrous earthquakes are not particularly rare. Every year there are several killer quakes. By contrast,
the seven catastrophes discussed in the preceding chapters are
predicted to descend to a catastrophic extent upon people all over
the world and are all exceedingly infrequent. Most have never
happened at all, others only once or twice. Still, many people live
in terror of earthquakes, earthquakes are regularly predicted by
scientists and nonscientific prophets alike, and the earthquake
Community competes aggressively with the exponents of other

243

predicted catastrophes for public attention and public funds. Hence, this chapter.

Earthquakes probably rank as the most terrifying of the many types of natural disasters that periodically beset mankind. They can and often do wreak large-scale damage, and they are almost always unexpected. They arrive with virtually no clear warning at any hour of the day, during any season of the year, and few locations on earth are secure from them—a sword hanging over our heads and liable to drop on us at any moment. Stories of cataclysmic earthquakes have found their way into the myths and religious scriptures of most civilizations. Major earthquakes have been carefully recorded since the earliest periods of written history in the Occident as well as in the Orient. For example, there is a catalogue of all of the substantial earthquakes that have been observed in central China over the past 3000 years. Impressively, complete historical records of quakes in Japan, Europe, and the Near East go back nearly 2000 years. These quakes have occurred regularly over the centuries and have consistently wrought great turmoil and suffering. Seers and savants capable (or claiming to be capable) of predicting the onset of earthquakes have always been in demand and have always received an inordinate amount of homage and attention.

In modern times, there are frequent and persistent reminders of the earthquake threat that hangs over us. These fall into three classes. First, many localities experience minor seismic tremors from time to time that are generally regarded as harbingers of bigger shocks to come. Second, on average, at least once each year we are shocked by reports of disastrous earthquakes and telecasts of cities in ruin and populations in great distress. In the past twenty years, there have been almost twenty earthquakes worldwide with death tolls exceeding 1000. Third, there are the frequent well-popularized predictions of catastrophic earthquakes to come. In this discussion, we are particularly concerned with the last of these classes—the prediction of the location, date, and intensity of major earthquakes.

An excellent and very readable scientific discussion of the

how and why of earthquakes is provided in Bruce Bolt's[2] *Earthquakes*, revised in 1988. Other readable books on the subject include John Nance's[3] *On Shaky Ground: America's Earthquake Alert* or David Ritchie's[4] *Superquake: Why Earthquakes Occur and When the Big One Will Hit*, both published in 1988.

THEORY

The Origin of Earthquakes. The theory that accounts satisfactorily for the tens of thousands of earthquakes that are observed each year is of relatively recent origin. Although there had been talk of "continental drift" since the early days of this century, the *tectonic plate* theory did not gain the acceptance of geologists until the late 1950s. According to this theory, approximately ten gigantic slabs of the earth crust, each about fifty miles thick, are floating on softer materials below. The continents ride on these slabs and move in a regular, predictable way. Continual flows of molten rock push upward from the earth's interior along well-defined ridges that run along the middle of the major oceans. This causes the plates, which ride on top of enormous convective currents in the plastic mantle, to move or to "spread" away from the midocean ridges. This movement in turn causes the plates to push each other and to crunch together.

Here and there the edge of one plate dips under the edge of an adjacent plate creating a so-called *subduction zone*. Where adjacent plates overlap, the oldest portions of the plates are submerged and return to the molten interior of the earth. The process has been likened to a set of conveyer belts that continuously and at a uniform rate carry material from the spreading zones to the subduction zones. The boundaries of the major tectonic plates, the spreading and subduction zones, and regions of great earthquake activity are shown in Figure 32.

The vast majority of earthquakes occur at the edges of the tectonic plates and are generated by the rubbing of one edge against the other. Pressure along the edges builds up over the

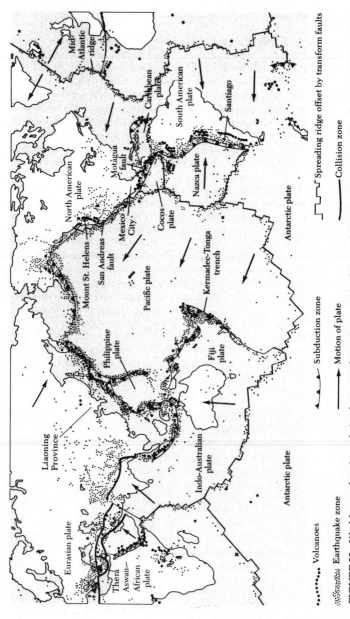

FIGURE 32. World map showing relation between the major tectonic plates and recent earthquakes and volcanoes. Earthquake epicenters are denoted by the small dots and the volcanoes by large dots (from *Earthquakes*, by Bruce A. Bolt,[5] copyright © 1978, 1988 by W. H. Freeman and Company, reprinted with permission).

Legend:
- •••••• Volcanoes
- ⠄⠄⠄⠄⠄ Earthquake zone
- ⊥⌐⊤⌐ Spreading ridge offset by transform faults
- ▲▲▲ Subduction zone
- → Motion of plate
- ⌐ Collision zone

Plate labels: Eurasian plate, Liaoning Province, Philippine plate, North American plate, Mount St. Helens, San Andreas fault, Mexico City, Pacific plate, Mid Atlantic ridge, Caribbean plate, Motagua fault, South American plate, Santiago, Nazca plate, Cocos plate, Kermadec-Tonga trench, Fiji plate, Indo-Australian plate, Thera, Aswan, African plate, Antarctic plate years, decades, and centuries until finally a rupture occurs at some point along a fault and becomes the *focus of an earthquake*: the point from which the first seismic waves emanate. Friction between the plates causes a gradual buildup of forces that are released irregularly, so that earthquake foci seem to appear sporadically in time and space.

Note that not all the earthquake zones shown in Figure 32 are located at the plate boundaries. There are also *intraplate earthquake zones*, which arise from other geologic conditions and include earthquake locations in the eastern United States and in the midwest. John Nance,[3] among others, argues that no place in America (or Europe) is safe from earthquakes. There are many faults and deep cracks, thousands in the continental crust that forms the North American plate. These may be thousands of feet below the surface and are well concealed by overlying geological formations, but become active and spread disaster over a wide region. Because of the composition of the crust, seismic shocks east of the Rockies can be expected to be transmitted over much greater distances than those triggered near the edges of the tectonic plates. For example, a major quake in the St. Louis area could well wreak havoc in such eastern cities as New York and Philadelphia. So the theory goes.

Characteristics of Earthquakes. Earthquakes give rise to a host of seismic waves, vibrations that are transmitted through the earth at speeds that depend upon the physical properties of the materials through which they travel. Hence, the vibrations that are felt are really a complicated combination of many waves. Their frequency components may vary from the low audio range (over 20 cycles per second) to resonance movements of only one cycle per hour. The most familiar indexes of earth movement, therefore, do not tell the whole story.

In Richter's method for characterizing the intensity of an earthquake, the maximum amplitude of the actual ground motion is measured. This value is then adjusted to take into account the distance between the point of measurement and the focus of the earthquake. The scale is logarithmic so that an increase of 1.0 on the Richter scale corresponds to an increase in amplitude by a factor of 10. For example, 6.0 characterizes a fairly strong earthquake, while, a 7.0 quake is ten times as strong and usually causes much more damage.

Another way of characterizing the magnitude of an earthquake is the Modified Mercali scale, which is based upon the damage caused to various buildings and other structures. The amount of destruction resulting from an earthquake is strongly dependent on local soil conditions, the type and quality of building construction, and, of course, the density of the population in the immediate area.

Prediction of Earthquakes. The approach to modeling used in simulating the atmosphere to predict the weather and the climate is not directly applicable to earthquake prediction. The tectonic plates are too remote from observation and too complicated to be described in great detail by a mathematical model. Computer simulations can be used to study the general movements of the plates and how the continents have drifted in the course of hundreds of millions of years. But there is no way to use this information to pinpoint the time and location of an earthquake. That information is lost in the uncertainty and noise associated with the model. It would be like asking a meteorologist to predict the precise minute and second that a rainstorm will start next week. For this reason, seismologists use a different technique to predict earthquakes. Their predictions are based on local observations that fall into two categories: precursors and historical records.

Since the dawn of history, humanity has tried to use the stars, magic, and a myriad of different "signs" to predict the time and place of major earthquakes. In fact, most of the forty-four means for foretelling the future, mentioned earlier, have been applied to earthquake prediction, and some of them are still in use today. Scientists, too, look for tell-tale signals, so-called "precursors" to provide advance warning of major quakes. Of course, the scientific precursors are based on some sort of cause/effect reasoning, on causality. Here are the categories of the on-site observations that may suggest the imminence of an earthquake:

Changes in Groundwater: Pressure changes in the earth immediately preceding an earthquake often cause unusual things to happen to the water in wells. The level of the water may fluctuate, and certain gases, such as radon, methane, and argon, may suddenly appear.

Electromagnetic Phenomena: When rocks are subjected to high pressure, they sometimes generate weak electric currents. These so-called *electrotelluric currents* can be measured and monitored.

Land Deformation: Earthquakes are sometimes preceded by uplifts and abrupt changes in the slope of the surface of the earth.

Seismicity: As the tectonic plates move against each other, a series of small earthquakes are generated. Frequently, just before a large earthquake, these microearthquakes diminish in magnitude and frequency.

Animal Behavior: Some animals appear to sense the imminence of an earthquake and, in response, exhibit all sorts of agitated or otherwise unusual patterns of behavior.

None of the above precursors appear often enough nor consistently enough to constitute reliable prediction mechanisms. Nonetheless, earthquake precursors are the subject of intensive study by scientists all over the world.

Seismic Gaps: Most earthquakes occur along fault lines at which geologic layers or strata come together or overlap. The regular and uniform movement of the tectonic plates causes continuous forces to be exerted along the faults that mark the plate boundaries. But because of friction, the plates do not slide along smoothly, rather they tend to lurch. Pressures build up and then cause sudden ruptures along the faults, and these in turn become earthquake foci. One way or another, every point along a plate boundary must move along. But at most locations the lurches are frequent and very small, and the earthquakes they produce are tiny. The longer the elapsed time between earthquakes, the more intense the eventual seismic shock. Therefore, an important approach to earthquake prediction involves the careful recording of all earthquakes, even microearthquakes along known faults and looking for significant *seismic gaps*—regions along the fault that have not had any recent earthquake activity of substantial magnitude. Those are the locations that are due or overdue for a major quake.

Periodicity: In some earthquake zones historical records show major earthquakes to recur at more or less regular intervals. For example, in central California the area near the town of Parkfield has experienced a fair-sized earthquake just about every twenty-two years since 1857. This has led to confident yet-to-be-fulfilled predictions of a quake in the 1987 to 1993 time frame. Some scientists have attempted to correlate the timing of these periodic quakes with other periodic phenomena such as earth tides, the location of the planets and the moon, and sunspots.

The problem with seismic gaps and periodic events is that they lead to predictions that are approximate at best. Time intervals of five to ten years are next to nothing on the geologic time scale. Most reputable scientists, therefore, couch their predictions in probabilistic terms. Some sophisticated statistical methods are available for this. They might conclude, for example, that there is a 45% probability of a 7.0 or greater earthquake in southern California in the next fifty years. Such a prediction is not useful for day-to-day planning, but it is valuable in drafting building codes for earthquake-safe structures and in making other long-range plans.

DAMAGE CONTROL

The hazards attending major earthquakes are numerous and diverse. The following are the major causes of fatalities immediately following an earthquake.

- Collapse of buildings and other structures due to ground shaking

- Collapse of structures due to tilting caused by ground settling
- Fires resulting from the destruction of buildings
- Avalanches and landslides
- Floods resulting by dam failures
- Tidal waves (tsunamis) due to earthquakes in the ocean

To these hazards must be added the long-term risks of famines, epidemics, and homelessness that result from massive physical destruction and the collapse of governmental infrastructure.

The many earthquakes taking place all over the world every year provide engineers with excellent opportunities to study all of the above hazards in great detail and to evaluate various approaches to limit loss of life and damage to property. A major problem in that regard is that the effect of major quakes on structures depends very much on local ground conditions—the proximity of bedrock, the presence of water, and so on. In general, however, engineers are capable of designing structures that will withstand earthquakes up to a specified magnitude. The greater that magnitude, the more expensive the necessary protective measures. By now, the building codes in many earthquake-prone regions include requirements for earthquake hardening of new structures. Here and there, extensive efforts are also under way to strengthen existing buildings and other structures to make them more earthquake resistant. Many municipalities also have put a variety of emergency procedures in place to minimize the confusion and panic that usually follow major quakes. All of these efforts are based on rather arbitrary assumptions as to the magnitude and character of expected earthquakes. They all entail judicious trade-offs of safety and cost.

DATA

The amount of available data regarding earthquakes and their effects is truly awesome. Major quakes leave indelible traces in rock formation, so there is geologic evidence of seismic shocks going back many millions of years. And, as already mentioned, great earthquakes appear prominently in the myths of most cultures and in the earliest historical writings. As we move into modern times, the records become more dense. For example, over 3600 earthquakes ranging from 3.0 to 6.9 in magnitude were recorded in northern and central California between 1949 and 1983. In addition, there is detailed information of the injuries and damage that accompanied each quake.

The "killer earthquakes" that lead to casualties of one thousand or more seem to occur about once a year on average. The most recent of these are listed in Table 11. The magnitude of these earthquakes ranges from 6.0 to 7.9, but there does not appear to be a direct correlation between the number of fatalities and the magnitude. The death toll is more affected by the population density near the focus and by local building practices. With few

TABLE 11.   Recent Earthquakes with over 1000 Fatalities
(for further information, see B. A. Bolt[6] and the World Almanac)

Date	Location	Magnitude	Death toll
June 1990	Iran	7.7	40,000
December 1988	Soviet Armenia	6.8	25,000
September 1985	Mexico	7.9	9,500
October 1983	Turkey	6.9	1,300
December 1982	Yemen	6.0	2,800
July 1981	Iran	7.3	1,500
June 1981	Iran	6.9	3,000
November 1980	Italy	7.2	4,800
October 1980	Algeria	7.7	25,000
March 1977	Romania	7.2	2,000
November 1976	Turkey	7.9	4,000
August 1976	Philippines	7.8	8,000
July 1976	China	7.6	242,000
February 1976	Turkey	6.8	2,300
December 1972	Nicaragua	6.2	5,000
May 1970	Peru	7.8	66,000

exceptions, the "killers" all occurred in the developing regions of the globe. In fact, we have to go back over 40 years to find an earthquake in an industrialized country that had fatalities exceeding five thousand. That happened in Japan in 1948 with 5100 casualties.

THE CAUSAL CHAIN

The fundamental cause of most earthquakes can be traced to movements of the tectonic plates forming the crust of the earth. Although formulated only relatively recently, the general model describing continental drift is well established and accepted. The tectonic plate theory accounts for the general pattern and location of most of the world's earthquakes. Well-defined earthquake zones generally lie along the boundaries of the plates. The theory is not useful, however, in predicting the time, location, and magnitude of specific earthquakes. The model does not lend itself to providing that kind of detail. Hence, earthquake predictions have to be based almost entirely on local observations, on careful readings of the historical record, and on drawing inferences from the various presumed precursors.

The precursors mentioned earlier in this chapter are not really part of the causal chain. They are effects that are caused by the same high pressures that lead to earthquakes. They are all reasonable from the scientific point of view, but they appear and disappear mysteriously. The effects of solar, lunar, and planetary alignments do produce large tidal forces at times and in retrospect can be regarded as triggers of earthquakes, but it is a bit like the straw that broke the camel's back. You can't tell in advance which straw will precipitate the catastrophe. Conclusions drawn from observations of seismic gaps and of repetitive patterns in past earthquakes likewise have a scientific justification, but there is little basis for predictions in other than broad probabilistic terms, as described below.

Therefore, with reference to the comprehensive model shown

in Figure 33, the prediction of the time, place, and magnitude of a specific earthquake, which is the output of Model C, is generated by the judicious mingling of three major components: the implications of seismic gaps, the presumed triggering effects of solar and lunar tides and of planetary alignments, and the diverse precursors. Each forecaster is likely to brew a different broth from these ingredients and therefore to come up with a different prediction. Some may regard all this as black magic, but in any event, the specific prediction of earthquakes falls into the very dark region of the spectrum of mathematical models discussed in Chapter 3.

The prediction of the casualties, the damage, and the long-term problems resulting from an earthquake, given its location and magnitude, is also fraught with many additional uncertainties. So much depends on the kinds of seismic waves that are generated, what minor or unsuspected faults are excited by the quake, how much moisture happens to be present in the soil, and so on. One building may crumble and collapse, while a neighboring building of the same construction may survive unscathed. One section of a freeway may collapse with great loss of life, while an adjacent section may remain totally unaffected. A 7.9 earth-

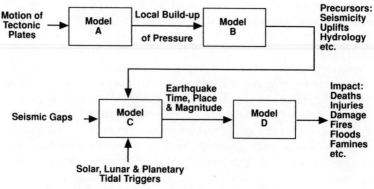

FIGURE 33. Comprehensive model of earthquakes.

quake in a populated area is sure to inflict very costly damage, but the death toll may vary from a few hundred to over 200,000. Hence, the output of Model D is far more uncertain than the already highly uncertain output of Model C.

PREDICTIONS

The killer quakes listed in Table 11 have one thing in common: not a single one of them was predicted scientifically. They all came completely unexpectedly. Looking at the past performance of scientists venturing earthquake predictions, we see precious few successes and many spectacular failures.

In February 1975, a group of Chinese scientists carefully examined a wide spectrum of precursors and concluded that a major earthquake was imminent near Haicheng in Manchuria. Warnings went out to the populace and most people spent the night outside, most domestic animals were removed from their stables, and expensive vehicles and other objects were moved out of garages and warehouses. Sure enough, a 7.3 quake promptly struck the region, causing 90% of the buildings to collapse, but very few of the three million people in the area were killed or injured. A great success. But only a year later, the same team of scientists completely failed to predict the 7.6 magnitude Tangshan earthquake that killed about a quarter of a million people. Precursors did not give unequivocal warnings of that quake.

In the late 1970s two major earthquake predictions in California incited many people there to push the panic button. In February 1976, a mysterious uplift of a large area near Palmdale, California was discovered, not far from the Los Angeles metropolitan area. The surface of the earth had risen anywhere from a few inches to as much as a foot. Many scientists viewed the so-called "Palmdale bulge" as a significant precursor and confidently predicted a quake in the 5.5 to 6.5 range before April 1977. At least 12,000 fatalities were anticipated. Eventually, the uplift went away and no killer quake materialized. Sociologist Ralph Turner[7] and his colleagues made a very detailed study of the manner in which news of the earthquake threat was disseminated and of the ways that the public responded to the warnings.

In 1974, John Gribbin and Stephen Plageman[8] published a book titled *The Jupiter Effect*, in which they presented massive scientific evidence that a major earthquake would hit California between 1977 and 1982, but most likely in mid-1982. The trigger of this superquake was to have been a very rare conjunction of the planets of our solar system. In 1982, the sun and all of the planets were positioned along a straight line, an event that takes place only once every 179 years. This alignment was expected to produce substantial tidal forces on the earth, which in turn would precipitate the earthquake. But, alas, the "big one" did not arrive. In the meantime, many people fled California, real estate prices plunged, and some people seriously believed that California would shortly be detached from the continental United States.

In 1974, 1975, and 1976, Brian T. Brady, a respected scientist working for the U.S. government, published a series of articles in scientific journals in which he described an earthquake prediction theory that combined the use of a number of precursors with physics and mathematics. He showed that this theory correctly "predicted," in retrospect, several of the earthquakes that had taken place earlier in the 1970s. He then went on to predict an earthquake, 9.2 in magnitude, to take place about fifty miles off the coast of Central Peru in the autumn of 1981. Early in 1981, Brady changed the date of the predicted earthquake several times and eventually settled on the June 28 to July 1, 1981 time frame. As described in detail by Richard Olson,[9] Brady's predictions received much publicity all over the world, but especially in the United States and, of course, in Peru. The Peruvian government took a number of emergency measures and U.S. and international agencies provided massive funds to mitigate the expected disaster. When the predicted dates passed uneventfully, there was a collective sigh of relief, but also a lot of hard feelings were evident.

In the late 1980s, Iben Browning, who has a Ph.D. in biology, but few credentials in geology or geophysics, predicted that an earthquake, 6.5 to 7.5 in magnitude, would strike the New Madrid area of Missouri, between December 1 and December 5, 1990. These predictions were based on a number of precursors as well as upon the expectation that an unusual alignment of the sun, moon, and earth would provide a tidal trigger for the quake. During the first week of December 1990, schools in southeastern Missouri were closed, numerous factories shut down, and many people fled the area altogether. Needless to say, the period of danger passed without a rumble.

Notwithstanding these disappointments, the earthquake prediction Community is alive and well. Major research efforts are under way in many parts of the world, and most scientists are confident that their ability to read precursors will gradually improve, making reliable earthquake predictions a reality. It is simply taking much longer than most people thought. In the meantime, there is no dearth of predictions. Here and there, a few scientists are venturing pinpoint predictions of date, place, and magnitude. But the less flamboyant majority is sticking to probabilistic forecasts. For example, a "blue ribbon" panel of scientists convened by the U.S. Geological Survey[10] in 1988 prognosticated the following:

- An earthquake 7.5 to 8.0 in magnitude along the San Andreas Fault near Los Angeles in the next thirty years with a 60% probability
- An earthquake with a magnitude of 6.5 to 7.0 along the San Jacinto fault in Southern California in the next thirty years with a probability of 50%
- A 7.0 magnitude earthquake in the San Francisco Bay area in the next thirty years with a probability of 50%.
- A 6.0 magnitude earthquake near Parkfield, California in the next five years with a probability greater than 90%

Predictions couched in similar terms are extant for most populated regions that lie near the boundaries of the tectonic plates. There are also forecasts of major earthquakes in places far from the plate boundaries—New York City, Charleston, South

Carolina, and Washington, D.C., among many others. Most often these prognostications are based on local precursors, such as swarms of small earthquakes.

WHAT SHOULD BE DONE

Virtually no scientist believes that major earthquakes can be prevented. The earthquake foci are too deep and too inaccessible for that. So the accent is entirely on prediction and damage control. The measures that are recommended to minimize fatalities, injuries, and property damage fall into two categories, depending on how much time is available until the arrival of a predicted major earthquake.

By now engineers have learned how to design and construct buildings and other structures to make them relatively safe from earthquakes up to a specified magnitude. In most developed countries, there are laws that assure that all new constructions include the necessary earthquake safeguards. These are applied with particular care to "sensitive" structures such as dams, bridges, nuclear reactors, hospitals, and schools. In regions where the earthquake probability is high, these safeguards are also imposed on industrial and office buildings and on private residences. This increases the cost of new construction, but is generally viewed as "good insurance" against a possible calamity. In a number of municipalities there are also extensive projects to "retrofit" some older buildings to make them more earthquake resistant and to condemn other buildings outright. These retrofit or condemnation measures can be very costly and inflict considerable hardship on those involved.

To cope with situations in which there is little or no advance warning of a large quake, many communities have adopted emergency preparatory programs. These include the training of government personnel and the formation of community groups to respond quickly and appropriately to an earthquake. At the same time, members of the public are being educated in how to equip the work place, schools, and homes to be ready for a quake, and how to act when the "big one" comes.

A PERSONAL VIEW

Among the catastrophes discussed here, earthquakes are unique in that their onset cannot be blamed on any human activity—not environmental pollution, not excessive energy consumption, not promiscuity, not even overpopulation. We don't know where or when the next one will strike, but we do know that its effects can be calamitous. Earthquakes seem to be initiated by cosmic forces, completely outside of our control. Earthquakes constitute a perpetual psychological burden, leading to a feeling of insecurity that haunts many of us all the time, a feeling that may well cause us to react emotionally rather than rationally. It is precisely because of the immediacy of the earthquake threat that we must be particularly cautious and circumspect in dealing with all scientific predictions of earthquakes.

I reject out of hand all predictions made by maverick scientists, in which they specify the more or less exact time, location, and magnitude of a large earthquake. I treat them the same way as the forecasts made by mystics, occultists, astrologers, and the like. They may entertain me, fascinate me, even scare me at times, but I would never consciously use them as the basis for any decision or action. It may be that a gifted individual, scientist or not, can have apparently astounding success as a forecaster, but until I see convincing evidence to the contrary, I would ascribe this feat entirely to luck. Credible models simply do not exist.

I am less negative about predictions that take the form of probabilities. But here, too, I have reservations—not so much in regard to the numerical probabilities as in regard to the way they are understood and used. As noted in Chapter 4, the field of probability is a respected mathematical discipline that is enormously useful in most of the sciences. However, even among scientists practicing the same specialty, controversies often break out because of the many ambiguities inherent in probabilistic specifications and because many scientists regard probabilities as being inherently subjective.

When it comes to communications with the public, the use of probabilities becomes even more tricky. The man in the street responds to predictions of two earthquakes, one with a probability of 90% and the other with a probability of 10%, in exactly the same way—by saying "Oh my God!" And he utters a sigh of relief when told that "there is a probability of 10% that there will not be an earthquake." This is because most people tend to confuse the expression "a probability of $X\%$" with the word "probable." Whatever the value assigned to X, they believe that the predicted event is likely to happen. As a result, I fear that most of us, even relatively sophisticated decision makers, are frequently misled by earthquake predictions. We tend to exaggerate the magnitude of the threat in our minds. Is this good or bad?

On one hand, it is good to make people aware of danger, it is good to nag and cajole them until they take the necessary remedial or mitigating actions. And even if the warning is a bit exaggerated, well, that will make them work all the harder. There is no such thing as being overly prepared for a catastrophe. Therefore, the more graphic, the more emphatic, the more terrifying the warning, the better. That's one way of looking at the issue.

But there are some compelling counterarguments. Most of them apply to all of the imminent catastrophes and are therefore discussed in detail elsewhere in this book. But earthquake catastrophes are different in the following respect. Each of the other seven catastrophes is global in character and calls for intervention at national and international levels, and invariably, there are exhortations for heavy governmental expenditures. In the United States, the federal government has very deep pockets. We can always find the money for a good cause; or at least we can pretend that we can. After all, look how much we spend for defense. Even the most ambitious programs can be made to appear to be modest in the general scheme of things.

By contrast, earthquakes are local catastrophes. They affect a very restricted area, not the country at large. Hence, many of the expenses involved in condemning unsafe buildings and in strengthening existing structures must be borne by local governments and local taxpayers. And their pockets are far from deep. Major expenditures must be balanced against each other. If one project receives more, another must receive less, which places local decision makers in a very difficult position. Suppose you were a mayor or a city manager and a group of scientists came to you and said: "The 'big one' is coming, and a lot of people are going to be killed. What are you going to do about it?" Visions of collapsed building and bridges, tens or even hundreds of thousands of constituents killed or injured, fires, floods, famines—a holocaust. As a political figure, could you possibly risk being blamed for failing to do enough should the earthquake actually come? How could you afford not to respond with open hands?

It would be very difficult to remain cool and logical under these circumstances. You might suspect that the funds requested for earthquake safety would be better spent for other purposes, projects that are almost certain to save more lives, even if the "big one" does come as predicted—improved police protection in high crime areas, campaigns to dissuade kids from smoking, medical check-ups, low-cost housing for the poor, and many others. None of these causes is able to point to a predicted sudden and catastrophic event to excite public interest. Poverty, crime, inadequate public health facilities, and urban neglect exact a steady toll in deaths and blighted lives that over a fifty-year period surely exceeds even the worst-case scenario of casualties in an earthquake by a large margin. How many more lives could be saved if earthquake funds were diverted in other directions? A difficult question, and one that needs to be approached in a rational manner, with a minimum of emotionalism.

Consider that in the past forty years, no earthquakes have resulted in five thousand fatalities in the developed parts of the world. Major earthquakes exact their toll primarily in parts of those regions where most people live in houses or hovels poorly constructed of masonry or stone, ready to collapse. It is very unlikely that a major trembler in a modern North American or European city would be fatal to more than five thousand inhabitants. This is not to say that five thousand fatalities anywhere would not be a major tragedy. But the prediction calls for a major earthquake sometime in the next fifty years, and with a probability of 50% at that. So these deaths should be regarded as spread over that period. In other words, we are really talking about a fraction of one hundred deaths per year, on average, in each of the next fifty years. If the primary objective is to save lives, it would seem that one hundred lives per year could be saved much more inexpensively by additional, small increases in expenditures for public safety, public health, and public welfare.

No, I am not opposed to making buildings more earthquake resistant. I believe in strict building codes. And I believe in preparedness. To me it is very much a question of degree. Many observers have likened expenditures for earthquake preparations, and indeed outlays for each of the imminent catastrophes, as a form of insurance. If lightning strikes, you will, of course, be glad that you bought the insurance policy. If not, at least you had the comfort of knowing that you were covered. But there is such a thing as being overly insured, especially when the premiums are very high. There comes a point beyond which expenditures for insurance become excessively burdensome and even counterproductive. This is obviously a difficult problem and one for which I have no pat solution to offer. I am reluctant to point a finger at anyone, for I am not exactly a paragon of rationalism in the way that I handle my personal finances.

I carry many different insurance policies: life insurance, automobile insurance, fire insurance, flood insurance, health insurance, accident insurance, and earthquake insurance. The earthquake insurance is definitely not a good buy. There is a $10,000 deductible, whereas most residential earthquake damage comes to less than that amount. There is a ceiling, which is too low to protect me completely; and the premium is very high, in part because in case of a big quake, the insurance company would be swamped with claims. So, I am probably throwing away money. Figuring cost/benefit ratios, I would be far better off with more life insurance, or perhaps with less insurance all around. And, yet, after making many calculations and thinking things over very carefully, I recently renewed the earthquake insurance on my home. Why? Because it makes me feel better.

PART III

What Does It All Mean?

THIRTEEN

The Big Picture

THE EIGHT BIG THREATS

What have we learned? As of 1990, we seem to be confronted with eight dreadful and dreaded catastrophes. Each of them is championed by a Community that feels that its catastrophe should get top billing. But most scientists in or out of these Communities agree that all eight constitute real dangers and that some sort of concerted action is urgently needed to combat all of them. I am in total agreement with this. Nothing that I say below should be interpreted as suggesting that we should sit on our hands, that we should limit ourselves to "research" and to wait. The question is what to do first.

In Table 12, I have juxtaposed the imminent catastrophes discussed in Part II. These are the threats described in the preceding eight chapters. Each of these chapters is based primarily on two or three recent books, each addressed to a general audience rather than to specialists. *These are all relatively authoritative books, written by influential authors and embody the considered opinions and judgments of the pertinent Communities. They all reflect concerned but conservative and unhysterical points of view.* The material in the last three columns of the table comes directly from these books. Far more radical and far more costly suggestions are offered in many

267

TABLE 12. Summary of Imminent Catastrophes

Catastrophic event	Main scientific area	Immediate cause	Predicted consequences	Proposed actions	Cost of proposed actions
Depletion of ozone layer	Atmospheric chemistry	Emission of CFCs	Skin cancers, eye damage, and other health problems; damage to plants/animals; changes in climate	Stop use of CFCs; find substitutes for CFCs	"Billions and billions of $"
Global warming	Climatology	Emission of carbon dioxide and other greenhouse gases	Climate modification; droughts, storms, rise of sea level	More research; reduce energy consumption; reduce use of fossil fuels; enhance forest growth	"5% of national defense budget"
Low-level radiation	Medicine	Nuclear reactor accidents; inadequate storage and disposal of nuclear waste	Increased birth defects, infant mortality, and cancers; damage to immune system	Phase out all nuclear reactors and build no new ones	"Trillions of $ in U.S. alone"

Acid rain	Biology	Emission of oxides of sulfur and nitrogen	Destruction of forests; damage to agricultural and aquatic animal and plant life	Reduce emission of oxides of sulfur and nitrogen; scrub exhausts; switch to alternative energy sources	"Many billions of $ in the U.S., Europe, and Japan"
AIDS epidemic	Medicine and epidemiology	Appearance of new viruses: HIV-1 and HIV-2	Rising number of incurable cases of AIDS; crushing burden on health-care facilities	More research; improved methods for prevention and treatment	"Trillions of $"
Population explosion	Demography	Uncontrolled birth rate; increased life spans	Famines; epidemics; riots; wars	Intensive birth control	No estimates
Worldwide economic depression	Economics	Unwise policies of U.S. and other governments	Collapse of banks, currencies, and entire economies; famines; riots; wars	Balanced budgets; control money supply	No estimates
Catastrophic earthquakes	Seismology	Tectonic plate motions; local geologic faults; poorly designed structures	Extensive fatalities; destructions of buildings; damage to transportation system	Extensive strengthening of existing buildings; stricter building codes	No estimates

of the books that are included in the bibliography and in the media's daily communications.

In constructing Table 12, I have resisted two temptations. I have not tried to add things up, to estimate how much it would cost if the recommendations of all eight Communities were adopted. Some might call that a cheap shot. After all, it is the general practice to request more funds than are absolutely needed and to settle gladly for a fraction of them. It would be a big mistake to let the sheer magnitude of the recommended tasks paralyze us into inaction.

I have also abstained from applying the Spectrum of Models, discussed in Chapters 3 and 4, to rank order the eight catastrophes. In fact, the predictions of all eight fall into the very dark, black end of the Spectrum of Models. The models used by the Communities may be quite valid and reliable when used to predict variables in their normal, well-studied range. Each of the catastrophe predictions, and actually all catastrophe predictions, are well out of those bounds. It follows that models and computer simulations cannot provide meaningful indications as to how and when any catastrophe will occur. Certainly there are catastrophes in our future, perhaps even fairly soon. But their magnitude and the point in time at which they will occur, even the year or decade, will come as a surprise, at least to scientists.

THE INTERRELATIONSHIP AMONG THREATS

Of course, the eight threats that we have discussed affect each other. They form part of the very complex fabric that characterizes the state of our global community and the many opportunities and pressures to which our decision makers must respond. As a result, coping effectively with one threat might make another threat even worse. For example, most improvements in public health can be expected to lengthen the average life span and exacerbate the population explosion. Reducing the use of fossil fuels so as to reduce the greenhouse effect may well encourage the increased use of nuclear energy with the attendant radiation hazards. And placing severe restrictions on the industries that are primarily responsible for acid rain may possibly hasten the onset of a catastrophic economic collapse. Hence, if our objective is somehow to optimize the quality of life for most of the people in the world, we cannot look at any one of the threats by itself. We must be careful to recognize and take into account the fact that all of the threats and their cures are interrelated. To predict the future, therefore, it is necessary to combine (the technical term is "link") all the models that have been designed to deal with specific problems. Unfortunately the "world models" that have been attempted from time to time have met with very little success.

The first major and ambitious attempt to model the world's future was undertaken in the early 1970s by a team of researchers at MIT.[1] Their book *The Limits to Growth* first appeared in 1972, went through three editions, and quickly became an international best-seller. The team, led by Dennis Meadows, used the systems dynamics approach to the modeling of complex systems devised by Professor Jay Forrester.

The basic model, termed "World 3," consists of five major components: resources, population, pollution, capital, and agriculture. Each of these is comprised of a number of subsystems, 121 in all. This model was run over a period of time from the year 1900 to the end of the twenty-first century. According to the book, "The standard model run assumes no major changes in the physical, economic or social relationships which have historically governed the development of the world system. All variables plotted here follow historical values from 1900 to 1970. Food, industrial output and population grow exponentially until the rapidly diminishing resource base forces a slowdown in industrial growth. Because of natural delays in the system, both population and pollution continue to increase for some time after the peak of industrialization. Population growth is finally halted by a rise in death rate due to decreased food and medical services."

The MIT team then explored how the model outputs would be affected when some of the basic assumptions were changed.

For example, they doubled the resource reserves assumed for 1900 while keeping all other assumptions identical to the standard run. This resulted in increased industrialization and pollution, while causing a very sharp (catastrophic) decline in world population around 2050. Another scenario assumed that nuclear power would provide "unlimited" resources. Sharp increases in pollution again led to a catastrophe in the middle of the twenty-first century. In fact, all attempts to provide "fixes," short of those recommended in *Limits to Growth*, were found to be of relatively minor benefit. Some even turned out to be counterproductive. The general conclusion of the study was that a continuing of existing trends in industrialization and population will result in a catastrophe—an abrupt decline in population of the order of five to seven billion in the next century, a calamity dwarfing in magnitude the casualties of all past wars, famines, and epidemics combined. The MIT team recommended drastic birth control and controls on industry and agriculture in order to achieve a satisfactory and stable "stationary state."

Some scientists[2,3] had favorable things to say about *Limits to Growth*, but many others were quite critical.[4,5] A book entitled *Models of Doom*, authored by a group of British researchers at the University of Sussex,[6] highlighted the weaknesses of the World 3 model: the questionable assumptions on which it is based, its tendency to exceed the bounds of validity of its submodels, and many other technical deficiencies. As a result, the Sussex researchers reasoned, the conclusions of *Limits to Growth* are unrealistic and of questionable validity. This sentiment was echoed by many others. For example, economists questioned the wisdom and practicality of placing limits on economic growth. Representatives of the Third World viewed the recommendations as part of a plot to inhibit their progress toward more political power and an enhanced standard of living.

Since the days of the *Limits to Growth* debate, there have been a number of other attempts to link models of the economy, population, industrialization, etc. in order to predict the future of the entire world or of specific regions such as Japan or Western

Europe. Some of these have been very useful to policy makers. An important study directed by Nobelist W. Leontief was sponsored by the United Nations. The Organization for Economic Cooperation and Development sponsored another widely quoted computer simulation effort. While many pessimistic scenarios have been explored using these linked models, none can be classified as out-and-out predictors of catastrophes.

IS THAT ALL? ARE WE AT A TURNING POINT?

Here are the essential questions we should ask: Are the eight catastrophes that we have discussed *it*? If somehow we prevail against them, can we sit back and enjoy life for a while?

To answer these questions, consider this scenario. Some thirty or more years ago, in a dense jungle in the heart of Africa, a little green monkey was sitting in a tree, minding its own business. Totally by accident a cosmic ray or perhaps a tiny radioactive particle passed through its body and produced a minute change in the genetic structure of a virus that had lived peacefully in generations of monkeys. Normally, such a mutant would be quickly eliminated by the monkey's immune system. But this time, the new virus thrived and multiplied. The little green monkey passed the virus on to other monkeys in the jungle, and after some years most of the green monkeys in that jungle were infected. Fortunately for them, the new virus did not bother them very much. Unfortunately for the human race, the virus was somehow passed on to some people dwelling in the jungle—unfortunately because the virus has the power of making humans very sick. In the years that followed, the virus was spread, first to people who came in contact with the natives of the jungle, then to other countries and continents, and, by the middle 1980s, we were faced with a worldwide AIDS epidemic.

This is, of course, a terribly unfortunate and very rare occurrence, a trillion to one shot perhaps, and quite unlikely to happen again. But remember this: every minute of every day, every monkey in the jungle and every other member of the animal kingdom is getting assaulted by cosmic rays and by other radiation that has the power to produce mutations, to give birth to new viruses. The chances of a new dread disease emerging as the result of a specific mutations is vanishingly small; but there are billions and billions of players in the game. Does it not seem likely that right now a number of plagues, as yet unheard of, are well on their way to fruition, to initiating another epidemic, another dreadful catastrophe? Of course it does. In fact, many scientists feel that we have been encouraging just such an eventuality by releasing radioactive matter into the atmosphere—through atomic bomb tests prior to 1963, by accidental releases from nuclear reactors, and by inadequate disposal of nuclear wastes. Right now we are terrified of AIDS. I wouldn't venture a prediction, but it stands to reason that another epidemic of some kind will be upon us sooner or later, probably sooner.

Also, consider this. In 1928, Thomas Midgley, Jr., a chemist working for Du Pont had a great idea. Refrigerators at that time used sulfur dioxide or ammonia as the coolant. They worked all right, but both sulfur dioxide and ammonia were poisonous to people and, therefore, unsafe to use. Refrigerators were destined to become a standard item in the households of the industrial nations; so something new and better was needed. Midgley invented a novel chemical compound consisting of carbon, fluorine, and chlorine, and other elements. The fluid that resulted was an excellent coolant, and it was neither poisonous nor inflammable. A very stable compound, it did not break down after time; it could be used indefinitely—a miracle chemical! Du Pont named it Freon.

By the 1930s, Freon dominated the refrigerator market, and, eventually, Freon and other similar chemicals, the so-called family of chlorofluorocarbons (CFCs), found many other industrial and consumer applications. Aerosols, in the familiar spray cans, became important products in hundreds of areas from cosmetics to insecticides—a thriving industry. Soon CFCs were lauded everywhere as a brilliant success of modern technology. They had the universal appeal of being both effective and harmless. As we have seen in Chapter 5, it was not until the mid-1980s, over fifty years later, that scientists became convinced that the CFCs were wreaking havoc on our atmosphere. It turned out that they served as catalysts in a chemical reaction by which ozone was annihilated in the outer atmosphere, creating a "hole in the sky," destroying what shielded us from dangerous ultraviolet light and the possibility of disastrous climate modifications.

We have in Freon and the CFCs a chemical that was hailed as an answer to the needs of humanity. People almost everywhere used it for over a half a century without anyone suspecting that it was responsible for an imminent catastrophe. In fact, it was more or less by accident that the threat posed by the CFCs was discovered. Now that threat is understood, and governments and industry are trying hard to do something about it, good!

But Freon is just one of hundreds and hundreds of "miracles" that the chemical industry has showered on us over the past decades: products that have been injected into the air that we breathe, the water that we drink, the food that we eat, the fuels that we put into our gas tanks, and on and on. What are the chances that, before long, one of these will be discovered to be deadly, to threaten to bring on another catastrophe? In fact, it would not be surprising if a number of such discoveries have already been made but have not as yet been publicized.

And do not overlook this. New faster and cheaper computers are appearing on the market every year. The amount of computing that can be done per dollar has been doubling every three or four years. Today's personal computers costing under $10,000 are already approaching the power of the first major supercomputers that cost seven or eight million dollars around twelve years ago. There seems to be no tapering off in this trend—wonderful! But the greater the computing power, the easier it becomes for scientists to implement larger and more complex models, to run bigger and bigger simulations of a wider and wider range of phenomena. We should, of course, welcome this. This is progress. But we must also recognize that as broader and broader ranges of phenomena are studied, as more and more scenarios are run on the computers, more potential catastrophes will be discovered—new catastrophes every bit as dangerous and every bit as imminent as the eight that we have discussed in Part II.

Fortunately, catastrophe threats not only burst upon us, but they also recede and disappear. The eight catastrophes are merely those that caught the forefront of the public consciousness around 1990. Ten or twenty years earlier, some of the eight were unknown and others were not considered to be terribly threatening. Overpopulation has been viewed as a major threat for a very long time; and so has the state of the economy. The dangers of radioactive fallout were appreciated, at least by some, in the early 1960s, and their activism led to the cessation of the atmospheric testing of atom bombs—over the objections of many scientists. But ten years ago, the ozone depletion and the AIDS threats had not surfaced yet, and more people were worried about global cooling than fretted about global warming due to the greenhouse effect.

Instead, other potential catastrophes were the focus of attention of scientists and of political activists. Since the 1950s, the threat of a nuclear holocaust, World War III, has been by far the most persistent of all nightmares. It was this perceived threat that prompted the escalation of the arms race and the expenditures of ruinous sums by the United States and the Soviet Union. Now that threat has receded.

Throughout most of the 1970s and well into the 1980s, the energy crisis was one of our principal concerns. The world was quickly running out of oil, and a catastrophe seemed imminent. The February 1981 issue of *National Geographic* was devoted to energy. It was assumed that we were soon to be without oil supplies. Conservative models and scenarios projected a price of $80 per barrel of oil by 1985. Crash programs to exploit shale and tar deposits were advocated along with nuclear fusion and a number of alternative sources of energy. Vast sums of money were spent on some of these as well as on nuclear fission reactors. Nineteen ninety brought another oil shock, but by then running out of energy was no longer considered to be an imminent threat.

What is waiting for us just around the corner? Maybe it is another disease, maybe another deadly pollutant. Perhaps a medicinal drug will turn out to have disastrous side effects; or nationalistic or religious fanaticism will make the top eight, or maybe something as yet undreamed of.

It seems likely, for example, that the pervasive electromagnetic fields that are radiated by power lines and electrical machinery will come to be considered a serious danger to the health of people living or working in their vicinity. In fact, a number of studies[7] have suggested that such people are at increased risk of developing cancer such as leukemia, lymph tumors, and brain tumors. Probably, sooner or later somebody will publish a popular book predicting an "electrical catastrophe," and that will compel decision makers in government and industry to engage in yet another painful reappraisal of the energy picture. On a different front, we already have books entitled *The Explosion of Terrorism*[8] and *Final Warning*,[9] warning us that "terrorism not nuclear war now pose the gravest threat to our society."

Hence, while major threats and problems remain unsolved, there is a continual turnover among those threats that are generally regarded as catastrophic or that get moved to the head of the line. What appears to be terribly urgent at one point in time diminishes in importance as new urgencies come along a few years later. We frequently hear exhortations such as: "This is our last chance!" "It is now or never!" "We are at a turning point in history!" It would appear that these sentiments reflect the human tendency to imbue the present moment with great importance and significance. In fact, the present moment is a turning point in history, as is every other moment, but only because the wheel is going round and round all the time.

It should be crystal clear by now that the infamous eight catastrophes are *not* all we need to watch for. We must accept the fact that a new, major, imminent catastrophe will come out of the woodwork every few years—a predicted catastrophe that will have the full backing of a reputable Community, that will get full attention from the media, that will arouse a full measure of political activism, and that will call for a prompt reordering of priorities. As our technology develops, as computers become more powerful, the breathing time between catastrophe predictions will become shorter and shorter. We must recognize that fact, and we must tune our response to catastrophe predictions to that fact.

FOURTEEN

Implications and Conclusions

The "eight imminent catastrophes" discussed in this book all represent "clear and present" threats to people everywhere on the globe. They are all cause for great concern, whether or not they actually grow to the catastrophic proportions predicted by scientific models. All merit attention and coordinated efforts to keep them from happening or at least minimize their impact.

No country, no government is doing enough to prevent any of the predicted catastrophes from happening. There is always the problem of money, of course, and the difficulty of getting agreements on the best course of action, agreements by those who represent divergent interests around the world. They include the rich, the poor, the religious, the captains of industry and their workers, the farmers, the city dwellers, the lovers of the great outdoors, and the list goes on and on. It is a challenge, but, surely, with enlightened and farsighted leadership and with the requisite sacrifices, progress is possible. It must be, because the stakes are so enormous.

How should we proceed? Should we focus on one or another catastrophe and concentrate all our money and efforts on it? If so, how do we choose? Or should we decide how much we want to spend to combat all of the catastrophes and somehow divide that amount among them? Is it now or never? Now or certain disaster?

We hear these questions raised all around us, and the calls for immediate action are becoming increasingly strident. How should we respond? We should strive to be rational, not to panic. In a cool, if not dispassionate manner, we must choose from the possible alternatives. But here we are very much handicapped by the way our society usually resolves critical issues.

THE HARMFUL NATURE OF ADVERSARIAL CONTESTS

Adversarial Encounters. Some of the institutions of our society are based directly or indirectly on adversarial confrontations. For example, in courts of law the litigants each choose an attorney, an advocate whose charge it is to present the case for his client in as favorable a light as possible. The advocate is not expected to be objective or impartial—quite the contrary. The underlying theory is that if both sides are well represented by dedicated partisans, justice and society are well served.

Similarly in sports, the athlete joins a team and is expected to do everything possible to defeat the opponent, within prescribed rules. Winning is what counts and nice guys finish last. Spectators cherish the adversarial confrontation of the two teams. They relish the starkness of the combat and the finality of the outcome. But there is more to it. An athletic contest is regarded as a microcosm, a morality play, symbolic of the serious, and sometimes lethal, struggles of the real world. Perhaps for this reason the jargon of athletic contests permeates everyday language. For example, in the business world we speak of "game plans," "getting to first base," "striking out," and "knockouts." The more aggressive the advocate, the greater is the admiration and reward that is reaped. Compromise and reflection are for wimps.

This adversarial approach may work well in certain areas. In most situations, however, this is definitely not the best way of doing things. In the course of their careers, scientists are often forced into adversarial situations. They compete for jobs and promotions, they compete to get their papers accepted by journals, and they compete for funds to finance their research projects. It can be argued that competition brings out the best efforts of the individuals involved and helps to sharpen the issues. But do we really want to use an adversarial system to cope with predicted catastrophes? That is what seems to be happening.

Behind each of the eight catastrophe predictions is a Community. Each Community wants to see laws passed and money appropriated so that it can wage the good fight first and foremost against *its* threat, be it ozone depletion, or the greenhouse effect, or acid rain, or AIDS, or whatever. All Communities are sincere. All realize that there will not be enough money, not enough activism, not enough political clout to take arms against all of the threats at once. So the Communities are forced into adversarial combat.

My Catastrophe Is More Catastrophic Than Your Catastrophe.
The competition between Communities takes a form quite different from that in a court of law where the advocates rail against each other while a trained judge keeps order. It is much more subtle. One Community rarely, if ever, makes overt moves to discredit another Community. The proponents of one catastrophe theory do not try to undermine the theory behind another cause. But there is stiff competition just the same. The media have only so much space to cover imminent disasters. Congressional committees have time for only so many hearings. There is a limit to the amount of money that governments can be induced to spend. Who prevails? The causes that manage to make the biggest waves get the most attention. Unfortunately, when the clamor erupts for these noble causes, the public has great trouble in deciding which of the catastrophe scenarios is the most critical and the most imminent.

Selecting which of the predicted catastrophes should get priority would be easy if we knew when they will occur and how devastating will be their impact. Unfortunately, scientists cannot give us a definitive answer, for that would require credible and valid models. In Chapter 4, I explained the factors that cause predictions to go astray: concepts such as "the curse of dimensionality" and "going out-of-bounds." The conclusion was that, by definition, most catastrophes are out-of-bounds. They have occurred so rarely, if they have occurred at all, that scientist lack the theory and the data necessary to construct valid models. The causal chains inevitably contain links that are constructed by pure guess work, inspired guess work perhaps, but guess work just the same. That is one of the sobering messages of this book.

Without reliable guidelines, the public and its elected representatives tend to react to catastrophe predictions in a frenetic and ad hoc fashion. When a new threat is discovered every two or three years, an assertive Community moves into the limelight. While it occupies center stage, the new catastrophe receives a measure of attention and resources, but rarely is it enough to make decisive progress, before it is pushed out of the way by a new catastrophe theory and a new Community. The upshot of the multiplicity of catastrophe predictions is that they tend to cancel each other out, leaving none with enough long-term support to do the job.

Imminent Catastrophes versus Secondary Threats. The catastrophes not only compete with each other, but with a host of other social urgencies that have not graduated to catastrophe status. Even in the most peaceful of times, society is beset by myriad crises and potential calamities, and its leaders and government agencies must strive to apportion their efforts and resources equitably. This is exceedingly difficult. It entails the careful consideration of trade-offs and compromises to satisfy the needs and interests of the many components of society. And it requires dealing with lobbyists and pressure groups.

When a Community goes public with a catastrophe prediction, it declares in effect: "Let me move to the head of the line of causes." The public, aroused and distracted by the imminent catastrophe, compels its public servants to take note; the government is *forced* into action; and social priorities are turned topsy-turvy. I will mention just a few of the causes that are likely to suffer.

At the top of the list are the services that governments provide for their citizens. For example, there is public safety. In 1991, over 24,000 Americans were victims of homicide. This is five times the annual death toll to be expected from skin cancer due to ozone layer depletion; and it is ten times the number of annualized worst-case fatalities expected to result from major earthquakes in the United States. If the primary objective is to save lives, expenditures for increased and improved police protection would appear to be far more cost effective than similar expenditures to mitigate some of the imminent catastrophes.

Then there is public and low-cost housing to alleviate the ever more virulent problems of the homeless and those living in urban squalor. The mental illnesses of the indigent represent an additional dimension and the direct or indirect cause of extreme miseries and fatalities. Public education, especially programs for the young and socially disadvantaged, desperately need more support. Beyond public services are causes to safeguard our environmental heritage. "Greens" in Europe and environmentalists in the United States generally support the Communities combating the greenhouse effect and acid rain, because these catastrophe-oriented Communities are committed to fighting for increased conservation of energy and the restriction of industrialization, urbanization, and deforestation. But these environmental causes are apt to be neglected if public support of the catastrophe Communities wanes.

The point is that when preference is given to the fight against well-publicized imminent catastrophes, other less glamorous but indisputably needy causes and projects are shortchanged. In the climate of media frenzy created by catastrophe predictions, the government's rationality often takes a back seat to political expediency.

THE NEED FOR MORE EFFECTIVE OVERSIGHT

Where We Stand. Each of the eight imminent catastrophes is terrifying in the havoc that it can wreak on our society. But every country looks at these threats differently and has different social priorities. In many parts of the globe, festering problems, such as strident nationalism, religious fanaticism, and racism, pose more immediate and more terrifying threats than the scientifically predicted catastrophes. Strong religious and cultural forces stand in the way of concerted actions. The widespread opposition to birth control is a case in point. As a result, in most countries the initiation of a large-scale attack against even a single threat requires Herculean effort, and, once launched, is subjected to all the vagaries of the political process. The United States is no exception.

In the United States, the struggle between Communities seeking support has taken on some of the aspects of an athletic tournament. However, it is a contest in which there are no clear rules and no final decisions. Each Community is going all out to convince the government and the public that it deserves the highest priority, while we depend on the murky political process to muddle to some sort of resolution. We need to rise above the parochial interests and desires of the Communities. We need to inject global perspectives and wisdom into the process.

The National Academy of Sciences, the National Academy of Engineering, and the Institute of Medicine were formed principally to serve as resources for the government, to provide counsel and advice on critical issues. The many federal departments and agencies make frequent and free use of them. There is, however, a glaring weakness in this arrangement, at least in how it is applied.

When a governmental body, such as the Department of Health and Welfare, needs advice on a pressing issue, it enlists the aid of the National Research Council to form a panel or committee of the best scientists. Although, on occasion, retirees are used, most of the panel members usually hold regular jobs in academia, industry, and the government and do their work on the panel as a pro bono, moonlighting contribution. They meet in Washington, D. C. for a few days every few months over a two- or three-year period and then return to their everyday activities.

The typical end product of each panel is a report dealing with a specific set of problems, as defined in the charge to the panel. Since the panel's role is purely advisory, the sponsoring body is free to accept or to ignore the advice it receives. As a result, many committee reports are gathering dust on book shelves filled with similar reports. So, while the government is able in this way to tap the best minds in the country, this process seldom gets results.

We need a permanent, highly professional organization to guide the government in dealing with predicted catastrophes and other major threats. You may say, "Oh no, not another Community." But I am serious; I'm convinced that we urgently need a mechanism to arbitrate between Communities, to determine the priorities to be assigned to existing threats and newly surfacing threats and to decide how resources should be allocated to fight them. We require government machinery to deal rationally with catastrophe management on a permanent, continuing basis rather than on a threat-to-threat basis. The following are some general suggestions to that end—by no means is this a full-fledged proposal.

The Board. I have in mind a government body (which I will call the Board), having some of the features of the National Science Foundation, which has a relatively free hand in expending substantial government funds to fulfill a specific mission. It should have some of the features of commissions that have broad powers to regulate industries, such as the Nuclear Regulatory Commission. Finally, it should be modeled on the Federal Reserve Board, which has the power and the means to exercise control over the U.S. economy by regulating the amount of money in circulation. The Federal Reserve Board is lobbied heavily by various advocacy communities representing competing interests: some favoring tight money, some lower interest rates, some lower taxes, some balanced budgets, some free trade, etc. Although not completely insulated from politics, the Federal Reserve is admired for being able to rise above the clamor of these advocates when it determines what is best for the country. This is a quality that is desperately needed in managing imminent catastrophes.

The Board might be run by a team of highly respected professionals, carefully and openly selected, nominated by the president and approved by Congress, with staggered terms of six or more years. The Board would be composed of senior scientists from the principal physical, life, and social science disciplines, of experts experienced in running large projects in the public and private sectors, as well as generalists knowledgeable in the workings of national and international political structures. The Board would act as a funnel for funds earmarked for catastrophe containment; it would be charged with budgeting the expenditures for the various critical causes; and it would commission studies and scientific investigations. It would also take over the coordination of the major national programs dealing with catastrophes that are already in operation, including those dealing with the greenhouse effect and acid rain.

The Board would also act as this country's interface with international organizations and programs concerned with specific catastrophes. Promising examples of far-reaching national and international efforts are the Global Change Research Program and the Ozone Protocols. It would be a charge of the Board to help nurture and expand such efforts and to help to assure that they receive long-term support sufficient to achieve realistic objectives.

GRACEFUL DEGRADATION

Most scientists discover early in their careers that there are problems in this world that do not have a solution, that many carefully stated objectives simply cannot be achieved because of the constraints imposed by mother nature and human nature. For example, as we have already mentioned, the chaos phenomenon precludes accurate long-range weather predictions, and the complexities of human communications have frustrated all attempts to perfect automatic translators of spoken and written languages.

Similarly, the Board may come to the realization that the achievement of an equilibrium in which the economy, the population, energy consumption, and pollution all have reached satisfactory levels is not in the cards. The Board must come to grips with constraints such as the religious and cultural opposition to rapid and effective population control and problems associated with the striving of all people for an improved standard of living and the amount of energy that must be generated to this end. These may lead the Board to conclude that objectives such as the conservation of our resources, the safeguarding of the environment, or the maintenance of our quality of life are impossible to achieve fully. In other words, a general deterioration of many of the things that we hold dear may be unavoidable.

Under these conditions, the Board may be led to borrow a page from an engineering textbook dealing with the design of complex systems that must function reliably for a long time. For example, when a spacecraft is sent on a multiyear voyage across the solar system, it is constantly peppered by stellar dust and meteorites. Most of these particles are too tiny to do any damage; they pass right through. Occasionally, though, the craft may be struck by a meteorite large enough to knock out an electronic component or to break an electrical connection. In old-fashioned, conventional designs, a single unlucky impact may suffice to disable the spacecraft completely, bringing to an abrupt end all the transmission of scientific measurement data and video images of distant planets. Since spacecraft are very precious, it is imperative to postpone such a minicatastrophe as long as possible.

Recognizing that meteorite impacts are random events from which there is no escape, engineers have adopted hardware and software techniques that greatly prolong the probable life of the spacecraft and its ability to perform at least some useful tasks. The key is the automatic restructuring of the spacecraft's circuits so that as some components are destroyed, others take their place in a manner that permits most systems to continue to work—only less efficiently.

It is a little like designing a car so that if a tire blows out, the spare tire is automatically inserted; then if another tire fails, the car is automatically converted into a tricycle; and when one of three remaining tires succumbs, the vehicle becomes a bicycle. In the case of the spacecraft, as more and more components and circuits are knocked out by meteorites, information is transmitted to earth more and more slowly, video images become less sharp, and there may be long dormant periods, until, finally, all is quiet and the spacecraft dies. Engineers have coined the term *graceful degradation* to characterize this interplay between life expectancy and the quality of life.

In applying the principle of graceful degradation, the Board and its counterparts around the world might well limit themselves to avoiding major and sudden catastrophic changes, while accepting the gradual deterioration of many valued qualities of the world we live in. Widespread starvation, epidemics, and civil strife may be obviated for a long time and economies kept on an even keel; but wilderness areas, jungles, and forests may gradually have to be reduced to a museum scale, natural resources may be used up, and pollution may become progressively more widespread.

This is obviously a grim prospect, but it may have a silver lining. Human values appear to adjust rapidly to circumstances. Suppose that a resident of present-day North America or Western Europe were magically given the chance to travel back in time to mid-nineteenth-century America. He or she would probably delight in the wide open spaces, forests, and prairies teeming with game, in the clear air and waters, in the absence of urban blight, and in the leisurely pace of everyday life. But our time traveler would very likely be shocked by the dismal state of medicine and sanitation, with the consequent low average life expectancy, and by the minimal opportunities for travel and education. The pervasive and glaring social injustices would probably also prove intolerable. In the end, the time traveler would most likely opt to leave those "good old days" and return to our own time and its more acceptable array of troubles.

But what about the future? Assume that somehow a total catastrophe, such as a nuclear holocaust, is avoided for the next century and a half and that a person born in the twenty-second century is transported to our own time. He or she would perhaps view our natural environment as pristine and idyllic, our cities as remarkably uncrowded, and our ways as quaint and wholesome. But my guess is that he or she would balance these pluses against the absence of a host of marvelous scientific, technological, and social innovations that we "old-timers" are unable to imagine or to appreciate, and would hurry back to the twenty-second century.

The point is that gradual change over time is inevitable. Whether the direction of this change is up or down, whether it is for better or for worse depends on the point of view and on the system of values. What seems like degradation in one era may be regarded as progress in another.

Most of the topics we have discussed are shrouded by unremitting gloom. This is to be expected in a book focused on the prediction of catastrophes. But we have been looking only at one side of the coin, the dark side. There is a host of prognosticators whose models predict good times ahead, with things improving exponentially: longer life spans, better health and welfare, prosperity, and general contentment. I have comparable reservations regarding those models, but that may be the subject of another book.

SOME FINAL THOUGHTS

Our society faces many serious challenges and threats. The struggle against the most imminent calamities, including the eight discussed in this book, must be carried out rationally and systematically.

Modern science provides us with valuable methods for modeling systems, for predicting their future course, and for evaluating proposed policies for coping with impending perils.

In order to make rational use of scientific models, decision makers and the public must acquire a broad understanding of the potentialities and the limitations of models. Only in this way can they attach the correct significance to predictions using them. There are no valid models for the prediction of the extraordinarily severe and rare calamities that we call catastrophes. We need a more effective mechanism for dealing on a continuing basis with all threats. In this book, I have suggested a broadly based Board, headed by a team of highly qualified scientists and generalists, to tackle these problems.

With wise and inspired leadership, we can chart and maintain the best course past the many imminent catastrophes. With a little luck, scientific and technological advances will continue to show us the way to a better world.

Mathematical Model of a Mechanical System

In this appendix we will look in more detail at the derivation of the mathematical model that characterizes the simple mass/spring system shown in Figure 7.

According to Newton's Laws, a mass exerts an inertial force that is the product of mass, M, times acceleration, a, and that if the mass exerts a downward force, the spring will exert an equal force in the upward (opposite) direction. Hence, at every instant of time the mass exerts a force, F:

$$F = Ma \qquad (A1)$$

while the spring exerts a force

$$F = -Ky \qquad (A2)$$

where the minus sign indicates that the spring is pulling up while the mass pulls down. We can now combine the two equations to read

$$Ma = -Ky \qquad (A3)$$

If we have studied a little calculus, we know that acceleration is the second derivative of displacement, so we can rearrange and rewrite Eq. (A3) and obtain

$$M(d^2y/dt^2) + Ky = 0 \qquad\qquad \text{(A4)}$$

where at $t = 0$, $y = -D$ and $dy/dt = 0$. The solution is a cosine wave,

$$y = -D\cos(2\pi ft) \qquad\qquad \text{(A5)}$$

where the frequency, f, in cycles per second, is

$$f = \frac{1}{2\pi}\sqrt{K/M} \qquad\qquad \text{(A6)}$$

So all we need to do is substitute the given values for K and M, and we have the desired solution.

If we include damping in the mathematical model, that is, if we assume that the spring dissipates energy, we must augment the model by including a damping coefficient, R, and an additional term in Eq. (A4), which now becomes

$$M(d^2y/dt^2) + R(dy/dt) + Ky = 0 \qquad\qquad \text{(A7)}$$

The solution now takes the form of a cosine wave, the amplitude of which decays exponentially with time.

References

CHAPTER 1

1. A. Bates, *Climates in Crisis* (The Book Publishing Company, Summertown, Tennessee, 1990).
2. R. Batra, *The Great Depression of 1990* (Venus Books, New York, 1985, 1987, 1988).
3. M. H. Brown, *The Toxic Cloud: The Poisoning of America's Air* (Harper & Row, New York, 1987).
4. R. Carson, *The Silent Spring* (Houghton Mifflin, Boston, 1962).
5. C. R. Chapman and D. Morrison, *Cosmic Catastrophes* (Plenum Press, New York, 1989).
6. P. R. Ehrlich and A. H. Ehrlich, *Extinction* (Ballantine Books, New York, 1981).
7. P. R. Ehrlich and A. H. Ehrlich, *The Population Explosion* (Simon and Schuster, New York, 1990).
8. L. Ephron, *The End: The Imminent Ice Age and How We Can Stop It* (Celestial Arts, Berkeley, 1988).
9. J. Fishman and R. Kalish, *Global Alert: The Ozone Pollution Crisis* (Plenum Press, New York, 1990).
10. P. G. Gale and T. Hauser, *Final Warning: The Legacy of Chernobyl* (Warner Books Inc., New York, 1988).

11. J. M. Gould and B. A. Goldman, *Deadly Deceit: Low Level Radiation, High Level Cover-up* (Four Walls Eight Windows, New York, 1988).
12. K. A. Gourlay, *Poisoners of the Seas* (ZED Books, New York, 1988).
13. J. Gribbin, *The Hole in the Sky* (Bantam, Corgi, New York, 1988).
14. J. Gribbin, *Hothouse Earth* (Grove Weidenfeld, New York, 1990).
15. K. J. Hsu, *The Great Dying: Cosmic Catastrophe, Dinosaurs, etc.* (Ballantine Books, New York, 1986).
16. T. Levenson, *Ice Time* (Harper & Row, New York, 1990).
17. F. Lyman, *The Greenhouse Trap* (Beacon Press, Boston, 1990).
18. B. McKibben *The End of Nature* (Random House, New York, 1980).
19. S. L. Roan, *Ozone Crisis* (Wiley, New York, 1989).
20. J. I. Slaff and J. K. Brubaker, *The AIDS Epidemic* (Warner Books, New York, 1985).
21. National Research Council, *Evaluation of Methodologies for Estimating Vulnerability to Electromagnetic Pulse Effects* (National Academy Press, Washington, D.C., 1984).
22. R. H. Turner, J. M. Nigg, and D. H. Paz, *Waiting for Disaster: Earthquake Watch in California* (University of California Press, Berkeley, 1986).

CHAPTER 2

1. R. E. Guiley, *The Encyclopedia of Witches and Witchcraft* (Facts on File, New York, 1989), pp. 103–105.
2. Editors of Time Magazine, *A Collection from Mysteries of the Unknown* (Time-Life Books, New York, 1989), p. 9.
3. D. Parker and J. Parker, *The Compleat Astrologer* (McGraw-Hill, New York, 1971).
4. R. Noorbergen, *Invitation to a Holocaust* (New English Library, London, 1981), p. 14.

5. S. Robb, *Prophecies on World Events by Nostradamus* (Liveright, New York, 1961), pp. 38, 123.
6. J. B. Payne, *Encyclopedia of Biblical Prophecy* (Harper & Row, New York, 1973), p. 596.
7. R. Noorbergen, *Invitation to a Holocaust* (New English Library, London, 1981), pp. 15–16.
8. N. Cohn, *The Pursuit of the Millennium* (Temple Smith, London, 1957).
9. M. Barkun, *Disaster and the Millennium* (Yale University Press, New Haven, 1974).

CHAPTER 3

1. T. S. Kuhn, *The Structure of Scientific Revolutions* (University of Chicago Press, Chicago, 1970).
2. W. J. Karplus, *The Spectrum of Mathematical Modeling and System Simulation*, Vol. 19 of *Mathematics and Computers in Simulation* (North-Holland, Amsterdam, 1977), pp. 3–10.

CHAPTER 4

1. K. C. Land and S. H. Schneider (eds.), *Forecasting in the Social and Natural Sciences* (Reidel, Boston, 1984).
2. H. W. Lewis, *Technological Risk* (Norton, New York, 1990).
3. J. Casti, *Searching for Certainty: What Scientists Can Know About the Future* (William Morrow, New York, 1990).
4. M. G. Morgan and M. Henrion, *Uncertainty: A Guide to Dealing with Uncertainty in Quantitative Risk and Policy Analysis* (Cambridge University Press, Cambridge and New York, 1990).

CHAPTER 5

1. S. L. Roan, *Ozone Crisis* (Wiley, New York, 1989).

2. J. Gribbin, *The Hole in the Sky* (Bantam, Corgi, New York, 1988).
3. J. Fishman and R. Kalish, *Global Alert: The Ozone Pollution Crisis* (Plenum Press, New York, 1990).

CHAPTER 6

1. A. Bates, *Climates in Crisis* (The Book Publishing Company, Summertown, Tennessee, 1990).
2. S. H. Schneider, *Global Warming* (Sierra Club Books, San Francisco, 1989).
3. J. Gribbin, *Hothouse Earth: The Greenhouse Effect and Gaia* (Grove Weidenfeld, New York, 1990).
4. S. H. Schneider, *Global Warming*, Ref. 2, p. 118.
5. S. H. Schneider, *Global Warming*, Ref. 2, p. 107.
6. S. H. Schneider, *Global Warming*, Ref. 2, p. 284.
7. L. Ephron, *The End: The Imminent Ice Age & How We Can Stop It* (Celestial Arts, Berkeley, 1988), p. 3.
8. L. Ephron, *The End: The Imminent Ice Age & How We Can Stop It*, Ref. 7.
9. D. E. Fisher, *Fire & Ice* (Harper & Row, New York, 1990).
10. J. Erickson, *Ice Ages: Past and Future* (Tab Books, Blue Ridge Summit, Pennsylvania, 1990).
11. L. Ephron, *The End: The Imminent Ice Age & How We Can Stop It*, Ref. 7, p. 131.
12. L. Ephron, *The End: The Imminent Ice Age & How We Can Stop It*, Ref. 7, p. 133.
13. L. Ephron, *The End: The Imminent Ice Age & How We Can Stop It*, Ref. 7, pp. 105–106.
14. D. E. Abrahamson (ed.), *The Challenge of Global Warming* (Island Press, Washington, D. C., 1989).
15. M. Allaby, *A Guide to Gaia: A Survey of the New Science of Our Living Earth* (Dutton, New York, 1990).
16. L. T. Edgerton, *The Rising Tide: Global Warming and World Sea Levels* (Island Press, Washington, D.C., 1991).

17. B. Erickson (ed.), *Call to Action: Handbook for Ecology, Peace and Justice* (Sierra Club Books, San Francisco, 1990).
18. J. Frior, *The Changing Atmosphere: A Global Challenge* (Yale University Press, New Haven, 1990).
19. H. P. Hymes, *Earthright* (Prima, New York, 1990).
20. J. Leggett (ed.), *Global Warming: The Greenpeace Report* (Oxford University Press, Oxford, 1990).
21. J. Naar, *Design for a Living Planet* (Harper & Row, New York, 1990).
22. J. J. Nance, *What Goes Up: The Global Assault on Our Atmosphere* (Morrow, New York, 1991).
23. G. Nult, *Cleaner, Safer, Greener* (Villard, New York, 1990).
24. M. Oppenheimer and R. H. Boyle, *Dead Heat: The Race Against the Greenhouse Effect* (Basic Books, New York, 1990).
25. R. Redford and T. J. Minger (eds.), *Greenhouse Glasnost* (Ecco Press, New York, 1990).
26. S. F. Singer, *Global Climate Change* (Paragon House, New York, 1989).
27. W. Steeger and J. Bowermaster, *Saving the Earth* (Knopf, New York, 1990).
28. J. Weiner, *The Next One Hundred Years* (Bantam Books, New York, 1990).

CHAPTER 7

1. J. M. Gould and B. A. Goldman, *Deadly Deceit: Low Level Radiation, High Level Cover-up* (Four Walls Eight Windows, New York, 1988), p. 1.
2. C. Sagan and R. Turco, *Path Where No Man Thought: Nuclear Winter and Its Implications* (Random House, New York, 1990).
3. J. M. Gould and B. A. Goldman, *Deadly Deceit: Low Level Radiation, High Level Cover-up*, Ref. 1.
4. P. G. Gale and T. Hauser, *Final Warning: The Legacy of Chernobyl* (Warner Books, New York, 1988).
5. D. Elsom, *Atmospheric Pollution* (Basil Blackwell, Oxford, 1987).

6. B. L. Cohen, *The Nuclear Energy Option* (Plenum Press, New York, 1990).
7. D. Elsom, *Atmospheric Pollution*, Ref. 5, p. 8.
8. J. M. Gould and B. A. Goldman, *Deadly Deceit: Low Level Radiation, High Level Cover-up*, Ref. 1, p. 107.
9. J. M. Gould and B. A. Goldman, *Deadly Deceit: Low Level Radiation, High Level Cover-up*, Ref. 1, pp. 39–70.

CHAPTER 8

1. C. C. Park, *Acid Rain: Rhetoric and Reality* (Methuen, New York, 1987), p. xii.
2. C. C. Park, *Acid Rain: Rhetoric and Reality*, Ref. 1.
3. G. Howells, *Acid Rain and Acid Waters* (Ellis Howard, New York, 1990).
4. D. D. Kemp, *Global Environmental Issues: A Climatological Approach* (Routledge, London, 1990).
5. D. Elsom, *Atmospheric Pollution* (Basil Blackwell, Oxford, 1987, 1989), p. 161.
6. C. C. Park, *Acid Rain: Rhetoric and Reality*, Ref. 1, p. 161.
7. C. C. Park, *Acid Rain: Rhetoric and Reality*, Ref. 1, p. 39.

CHAPTER 9

1. A. Friedman-Kien, in J. I. Slaff and J. K. Brubaker, *The AIDS Epidemic* (Warner Books, New York, 1985), p. 3.
2. J. I. Slaff and J. K. Brubaker, *The AIDS Epidemic*, Ref. 1.
3. P. O'Malley (ed.), *The AIDS Epidemic: Private Rights and Public Interest* (Beacon Press, Boston, 1988).
4. Editors of Scientific American Readings, *The Science of AIDS* (W. H. Freeman, New York, 1989).
5. M. Fumento, *The Myth of Heterosexual AIDS* (Basic Books, New York, 1990).

6. B. Nussbaum, *Good Intentions: How Big Business and the Medical Establishment Are Corrupting the Fight Against AIDS* (The Atlantic Monthly Press, New York, 1990).

7. R. R. Redfield and D. S. Burke, in *The Science of AIDS*, Ref. 4, pp. 63–74.

8. *The Science of AIDS*, Ref. 4, pp. 51–61.

9. M. A. Sande and P. A. Volberding, *The Medical Management of AIDS*, Second Edition (W. B. Saunders, Philadelphia, 1990), p. 40.

10. M. A. Sande and P. A. Volberding, *The Medical Management of AIDS*, Ref. 9, p. 5.

11. M. A. Sande and P. A. Volberding, *The Medical Management of AIDS*, Ref. 9, pp. 404–405.

12. M. A. Sande and P. A. Volberding, *The Medical Management of AIDS*, Ref. 9, p. 7.

13. V. deGruttola and W. I. Bennet, in *The AIDS Epidemic: Private Rights and Public Interest*, Ref. 3, pp. 152–153.

14. J. M. Mann *et al.*, *The Science of Aids*, Ref. 4, p. 51.

15. L. O. Gostin (ed.), *AIDS and the Health Care System* (Yale University Press, New Haven, 1990), pp. 197–210.

16. T. J. Matthews and D. P. Bolognesi, in *The Science of AIDS*, Ref. 4, p. 101.

17. R. A. Berk (ed.), *The Social Impact of AIDS in the U.S.* (Abt Books, Cambridge, 1988).

18. J. Dixon, *Catastrophic Rights: Experimental Drugs and AIDS* (New Star Books, Vancouver, 1990).

19. E. Fee and D. M. Fox (eds.), *AIDS: Burdens of History* (University of California Press, Berkeley, 1988).

20. M. D. Grmek, *History of AIDS: Emergence and Origin of a Modern Pandemic* (Princeton University Press, Princeton, 1990).

21. E. Kubler-Ross, *AIDS: The Ultimate Challenge* (Macmillan, New York, 1987).

22. C. Perrow and M. F. Guillen, *The AIDS Disaster: The Failure of Organizations in New York and the Nation* (Yale University Press, New Haven, 1990).

23. I. B. Corliss and M. Pittman-Lindeman (eds.), *AIDS: Principles, Practices & Politics* (Hemisphere, New York, 1988).
24. H. G. Miller, C. F. Turner, and L. E. Moses (eds.), *AIDS: The Second Decade* (National Academy Press, Washington, D.C., 1990).

CHAPTER 10

1. P. R. Ehrlich and A. H. Ehrlich, *The Population Explosion* (Simon and Schuster, New York, 1990), p. 3.
2. P. R. Ehrlich, *The Population Bomb* (Ballantine Books, New York, 1978).
3. P. R. Ehrlich and A. H. Ehrlich, *The Population Explosion*, Ref. 1.
4. P. R. Ehrlich and A. H. Ehrlich, *Extinction* (Ballantine Books, New York, 1981).
5. D. Winch, *Malthus* (Oxford University Press, Oxford, 1987).
6. P. R. Ehrlich and J. P. Holdren, "Impact of Population Growth," *Science* **171**, 1212–1217 (1971).
7. P. R. Ehrlich and A. H. Ehrlich, *Extinction*, Ref. 4, pp. 123–212.
8. J. L. Simon and H. Kahn (eds.), *The Resourceful Earth: A Response to Global 2000* (Basil Blackwell, New York, 1984).
9. *World Population: Fundamentals of Growth* (Population Reference Bureau, Washington, D.C., 1984).
10. T. W. Merrick *et al.*, "World Population in Transition," *Pop. Bull.* **41**(2), 8–12 (1989).
11. *Population Handbook*, Second Edition (Population Reference Bureau, Washington, D. C., 1990), pp. 56–63.
12. T. W. Merrick *et al.*, "World Population in Transition," *Pop. Bull.* Ref. 10, 12–13.
13. *World Population Data Sheet* (Population Reference Bureau, Washington, D. C., 1990).
14. *Population Handbook*, Second Edition, Ref. 11, p. 14.
15. T. W. Merrick *et al.*, "World Population in Transition," *Pop. Bull.* Ref. 10, 12–15.

16. *World Population Data Sheet*, Ref. 12.
17. L. Bouvier, "Planet Earth 1984–2034: A Demographic Vision," *Pop. Bull.* **39**(1) (1984).

CHAPTER 11

1. R. Batra, *The Great Depression of 1990* (Dell, New York, 1988), p. 9.
2. J. L. King, *How to Profit from the Next Great Depression* (Signet, New York, 1988), p. 19.
3. H. J. Ruff, *How to Prosper During the Coming Bad Years* (Warner Books, New York, 1979).
4. D. R. Casey, *Crisis Investing: Opportunities and Profits in the Coming Great Depression* (Harper & Row, New York, 1980).
5. J. F. Smith, *The Coming Currency Collapse and What You Can Do About It* (Books in Focus, New York, 1980).
6. J. Granville, *The Warning: The Coming Great Crash in the Stock Market* (Freundlich Books, New York, 1985).
7. P. Jay and M. Stewart, *Apocalypse 2000: Economic Breakdown and the Suicide of Democracy* (Prentice-Hall, New York, 1987).
8. P. Erdman, *What's Next? How to Prepare Yourself for the Crash of '89* (Doubleday, New York, 1988).
9. D. McClain, *Apocalypse on Wall Street* (Dow Jones Irwin, Homewood, Illinois, 1988).
10. R. Batra, *Surviving the Great Depression of 1990* (Simon and Schuster, New York, 1988).
11. H. Browne, *The Economic Time Bomb: How You Can Profit from the Emerging Crisis* (St. Martin's Press, New York, 1989).
12. C. W. J. Granger, *Forecasting in Business and Economics*, Second Edition (Academic Press, Boston, 1989).
13. C. P. Kindleberger, *Manics, Panics, and Crashes* (Basic Books, New York, 1989).
14. R. Batra, *Surviving the Great Depression of 1990*, Ref. 10, pp. 213–233.

15. H. Kahn, *The Coming Boom* (Simon and Schuster, New York, 1982).
16. R. Sobel, *Panic on Wall Street* (Dutton, New York, 1968, 1988).

CHAPTER 12

1. D. Ritchie, *Superquake: Why Earthquakes Occur and When the Big One Will Hit* (Crown, New York, 1988), p. 1.
2. B. A. Bolt, *Earthquakes*, Revised Edition (W. H. Freeman, New York, 1988).
3. J. J. Nance, *On Shaky Ground: America's Earthquake Alert* (William Morrow, New York, 1988).
4. D. Ritchie, *Superquake: Why Earthquakes Occur and When the Big One Will Hit*, Ref. 1.
5. B. A. Bolt, *Earthquakes*, Ref. 2, p. 13.
6. B. A. Bolt, *Earthquakes*, Ref. 2, pp. 229–238.
7. R. H. Turner, J. M. Nigg, and D. H. Paz, *Waiting for Disaster: Earthquake Watch in California* (University of California Press, Los Angeles, 1986).
8. J. R. Gribbin and S. H. Plageman, *The Jupiter Effect* (Walker, New York, 1974).
9. R. S. Olson, *The Politics of Earthquake Prediction* (Princeton University Press, Princeton, 1989).
10. "Probabilities of Large Earthquakes Occurring in California on the San Andreas Fault," U. S. Geologic Survey Report No. 88-398 (Federal Center, Denver, Colorado, 1988).

CHAPTER 13

1. D. W. Meadows, D. L. Meadows, J. Randers, and W. W. Behrens, *The Limits to Growth* (Universe Books, New York, 1972).
2. M. Mesarovic and E. Pestel, *Mankind at the Turning Point* (E. P. Dutton, New York, 1974).

3. L. J. Perelman, *The Global Mind: Beyond the Limits to Growth* (Mason/Charter, New York, 1976).
4. B. B. Hughes, *World Futures: A Critical Analysis of Alternatives* (The Johns Hopkins University Press, Baltimore, 1985).
5. R. McCutcheon, *Limits of the Modern World: A Study of the Limits to Growth Report* (Butterworths, London, 1979).
6. H. S. D. Cole, C. Freeman, M. Jahoda, and K. R. L. Pavitt, *Models of Doom* (Universe Books, New York, 1973).
7. "Electromagnetic Fields: The Jury's Still Out," *IEEE Spectrum*, August, 1990, 22–35.
8. B. Grosscup, *The Explosion of Terrorism* (New Horizon Press, Farhill, New Jersey, 1987).
9. R. H. Kupperman and J. Kamen, *Final Warning: Averting Disaster in the New Age of Terrorism* (Doubleday, New York, 1989).

Bibliography

Abrahamson, D. E. (ed)., *The Challenge of Global Warming* (Island Press, Washington, D.C., 1989).

Allaby, M., *A Guide to Gaia: A Survey of the New Science of Our Living Earth* (Dutton, New York, 1990).

Barkun, M., *Disaster and the Millennium* (Yale University Press, New Haven, 1974).

Bates, A., *Climates in Crisis* (The Book Publishing Company, Summertown, Tennessee, 1990).

Batra, R., *The Great Depression of 1990* (Venus Books, New York, 1985, 1987, 1988).

Batra, R., *Surviving the Great Depression of 1990* (Simon and Schuster, New York, 1988).

Berk, R. A. (ed.), *The Social Impact of AIDS in the U.S.* (Abt Books, Cambridge, Massachusetts, 1988).

Bolt, B. A., *Earthquakes* (W. H. Freeman, New York, 1978).

Bouvier, L. F., "Planet Earth 1984–2034: A Demographic Vision," *Pop. Bull.* **39**(1) (1984).

Brown, M. H., *The Toxic Cloud: The Poisoning of America's Air* (Harper & Row, New York, 1987).

Brown, P., and E. J. Mikkelsen, *No Safe Place: Toxic Waste, Leukemia and Community Action* (University of California Press, Berkeley, 1990).

Browne, H., *The Economic Time Bomb* (St. Martin's Press, New York, 1989).

Calder, N., *The Comet Is Coming* (Viking Press, New York, 1981).

Carson, R., *The Silent Spring* (Houghton Mifflin, Boston, 1962).

Casey, D. R., *Crisis Investing: Opportunities and Profits in the Coming Great Depression* (Harper & Row, New York, 1980).

Casti, J. L., *Searching for Certainty: What Scientists Can Know About the Future* (William Morrow, New York, 1990).

Chapman, C. R., and D. Morrison, *Cosmic Catastrophes* (Plenum Press, New York, 1989).

Close, F. E., *Apocalypse When? Cosmic Catastrophe and the Fate of the Universe* (William Morrow, New York, 1988).

Cohen, B. L., *The Nuclear Energy Option* (Plenum Press, New York, 1990).

Cohn, N., *The Pursuit of the Millennium* (Temple Smith, London, 1970).

Cole, H. S. D., C. Freeman, M. Jahoda, and K. R. L. Pavitt, *Models of Doom* (Universe Books, New York, 1973).

Corliss, I. B., and M. Pittman-Lindeman (eds.), *AIDS: Principles, Practices & Politics* (Hemisphere, New York, 1988).

Day, D., *The Environmental Wars* (St. Martin's Press, New York, 1989).

de Fontbrun, J.-C., *Nostradamus: Countdown to Apocalypse* (Holt, Rinehart & Winston, New York, 1983).

Dixon, J., *Catastrophic Rights: Experimental Drugs and AIDS* (New Star Books, Vancouver, 1990).

Dublin, M., *Futurehype: The Tyranny of Prophesy* (Dutton, New York, 1991).

Edgerton, L. T., *The Rising Tide: Global Warming and World Sea Levels* (Island Press, Washington, D. C., 1991).

Ehrlich, P. R., *The Population Bomb* (Ballantine Books, New York, 1968, 1971, 1978).

Ehrlich, P. R., and A. H. Ehrlich, *Extinction* (Ballantine Books, New York, 1981).

Ehrlich, P. R., and A. H. Ehrlich, *The Population Explosion* (Simon and Schuster, New York, 1990).

Elsom, D., *Atmospheric Pollution* (Basil Blackwell, Oxford, England, 1987, 1989).

Ephron, L., *The End: The Imminent Ice Age & How We Can Stop It* (Celestial Arts, Berkeley, 1988).

Erdman, P., *What's Next? How to Prepare Yourself for the Crash of '89* (Doubleday, New York, 1988).

Erickson, B. (ed.), *Call to Action: Handbook for Ecology, Peace & Justice* (Sierra Club Books, San Francisco, 1990).

Erickson, J., *Ice Ages: Past and Future* (Tab Books, Blue Ridge Summit, Pennsylvania, 1990).

Erikson, J., *Greenhouse Earth: Tomorrow's Disaster Today* (Tab Books, Blue Ridge Summit, Pennsylvania, 1990).

Fee, E., and D. M. Fox (eds.), *AIDS: Burdens of History* (University of California Press, Berkeley, 1988).

Fisher, D. E., *Fire & Ice* (Harper & Row, New York, 1990).

Fishman, J., and R. Kalish, *Global Alert: The Ozone Pollution Crisis* (Plenum Press, New York, 1990).

Foreman, D., *Confessions of an Eco-Warrior* (Harmony Books, New York, 1991).

Forman, H. J., *The Story of Prophecy* (Tudor, Greensboro, North Carolina, 1940).

Friedrich, O., *The End of the World* (Coward, McCann & Geohegan, New York, 1982).

Frior, J., *The Changing Atmosphere: A Global Challenge* (Yale University Press, New Haven, 1990).

Fumento, M., *The Myth of Heterosexual AIDS* (Basic Books, New York, 1990).

Gale, P. G., and T. Hauser, *Final Warning: The Legacy of Chernobyl* (Warner Books, New York, 1988).

Glickman, T. S., and M. Gough (eds.), *Readings in Risk* (Resources for the Future, Washington, D.C., 1990).

Gordon, A., and D. Suzuki, *It's a Matter of Survival* (Harvard University Press, Cambridge, Massachusetts, 1990).

Gordon, S. I., *Computer Models in Environmental Planning* (Van Nostrand-Reinhold, New York, 1985).

Gostin, L. O. (ed.), *AIDS and the Health Care System* (Yale University Press, New Haven, 1990).

Gould, J. M., and B. A. Goldman, *Deadly Deceit: Low Level Radiation, High Level Cover-up* (Four Walls Eight Windows, New York, 1988).

Gourlay, K. A., *Poisoners of the Seas* (ZED Books, New York, 1988).

Granger, C. W. J., *Forecasting in Business and Economics*, Second Edition (Academic Press, Boston, 1989).

Granville, J., *The Warning: The Coming Great Crash in the Stock Market* (Freundlich Books, New York, 1985).

Gribbin, J., *The Hole in the Sky* (Bantam, Corgi, New York, 1988).

Gribbin, J., *Hothouse Earth: The Greenhouse Effect and Gaia* (Grove Weidenfeld, New York, 1990).

Grmek, M. D., *History of AIDS: Emergence and Origin of a Modern Pandemic* (Princeton University Press, Princeton, New Jersey, 1990).

Grosscup, B., *The Explosion of Terrorism* (New Horizon Press, Farhill, New Jersey, 1987).

Guiley, R. E., *The Encyclopedia of Witches and Witchcraft* (Facts on File, New York, 1989).

Hall, M. P., *Secret Teachings of All Ages: An Encyclopedic Outline of Masonic, Hermetic, Qabbalistic and Rosecrucian Symbolical Philosophy*, Tenth Edition (The Philosophical Research Society, Los Angeles, 1952).

Haskell, T. L. (ed.), *The Authority of Experts* (Indiana University Press, Bloomington, 1984).

Hogue, J., *Nostradamus & the Millennium: Predictions of the Future* (Bantam-Doubleday-Dell, New York, 1987).

Holing, D., *Coastal Alert: Ecosystems, Energy, and Offshore Oil Drilling* (Island Press, Washington, D.C., 1990).

Howells, G., *Acid Rain and Acid Waters* (Ellis Horwood, New York, 1990).

Hughes, B. B., *World Futures: A Critical Analysis of Alternatives* (The Johns Hopkins University Press, Baltimore, 1985).

Hsu, K. J., *The Great Dying* (Ballantine Books, New York, 1986).

Hymes, H. P., *Earthright* (Prima, New York, 1990).

Ing, D., *The Chernobyl Syndrome* (Baen Books, New York, 1988).

Jagger, J., *The Nuclear Lion* (Plenum Press, New York, 1991).

Jay, P., and M. Stewart, *Apocalypse 2000: Economic Breakdown and the Suicide of Democracy* (Prentice-Hall, New York, 1987).

Kahn, H., *The Coming Boom* (Simon and Schuster, New York, 1982).

Karplus, W. J., *Analog Simulation: Solution of Field Problems* (McGraw-Hill Book Company, New York, 1958).

Karplus, W. J., *The Spectrum of Mathematical Modeling and System Simulation*, Vol. 19 of *Mathematics and Computers in Simulation* (North-Holland, Amsterdam, 1977), pp. 3–10.

Karplus, W. J., and V. Vemuri, *Digital Computer Treatment of Partial Differential Equations* (Prentice-Hall, New York, 1981).

Kemp, D. D., *Global Environmental Issues: A Climatological Approach* (Routledge, London, 1990).

Kindleberger, C. P., *Manics, Panics, and Crashes* (Basic Books, New York, 1989).

King, J. L., *How to Profit from the Next Great Depression* (Signet, New York, 1988).

Kingsland, S. E., *Modeling Nature* (University of Chicago Press, Chicago, 1985).

Kubler-Ross, E., *AIDS: The Ultimate Challenge* (Macmillan, New York, 1987).

Kuhn, T. S., *The Structure of Scientific Revolutions*, Second Edition (University of Chicago Press, Chicago, 1970).

Kupperman, R. H., and J. Kamen, *Final Warning: Averting Disaster in the New Age of Terrorism* (Doubleday, New York, 1989).

Land, K. C., and S. H. Schneider (eds.), *Forecasting in the Social Sciences* (Reidel, Boston, 1984).

Leggett, J. (ed.), *Global Warming: The Greenpeace Report* (Oxford University Press, Oxford, 1990).

Levenson, T., *Ice Time* (Harper & Row, New York, 1990).

Lewis, H. W., *Technological Risk* (Norton, New York, 1990).

Lewinsohn, R., *Prophets and Predictions: The History of Prophesy from Babylon to Wall Street* (Secker & Warburg, London, 1958).

Lovelock, J. E., *Gaia: A New Look at Life on Earth* (Oxford University Press, Oxford, 1979).

Lyman, F., *The Greenhouse Trap* (Beacon Press, Boston, 1990).

McClain, D., *Apocalypse on Wall Street* (Dow Jones Irwin, Homewood, Illinois, 1988).

McCutcheon, R., *Limits of the Modern World: A Study of the Limits to Growth Report* (Butterworths, London, 1979).

McKibben, B., *The End of Nature* (Random House, New York, 1989).

Meadows, D. H., D. L. Meadows, J. Randers, and W. W. Behrens, *The Limits to Growth* (Universe Books, New York, 1972).

Merrick, T. W., *et al.*, "World Population in Transition," *Pop. Bull.* **41**(2) (1989).

Mesarovic, M., and E. Pestel, *Mankind at the Turning Point* (Dutton, New York, 1974).

Miller, H. G., C. F. Turner, and L. E. Moses (eds.), *AIDS: The Second Decade* (National Academy Press, Washington, D. C., 1990).

Morgan, M. G., and M. Henrion, *Uncertainty: A Guide to Dealing with Uncertainty in Quantitative Risk and Policy Analysis* (Cambridge University Press, Cambridge, 1990).

Morine, D. E., *Good Dirt: Confessions of a Conservationist*. (The Globe Pequot Press, Chester, Connecticut, 1990).

Naar, J., *Design for a Living Planet* (Harper & Row, New York, 1990).

Nance, J. J., *On Shaky Ground: America's Earthquake Alert* (William Morrow, New York, 1988).

Nance, J. J., *What Goes Up: The Global Assault on Our Atmosphere* (William Morrow, New York, 1991).

National Research Council, *Valuing Health Risks, Costs, and Benefits for Environmental Decision Making* (National Academy Press, Washington, D. C., 1990).

Noorbergen, R., *Invitation to a Holocaust* (New English Library, London, 1981).

Nult, G., *Cleaner, Safer, Greener* (Villard, New York, 1990).

Nussbaum, B., *Good Intentions: How Big Business and the Medical Establishment Are Corrupting the Fight Against AIDS* (The Atlantic Monthly Press, New York, 1990).

O'Malley, P. (ed.), *The AIDS Epidemic: Private Rights and Public Interest* (Beacon Press, Boston, 1988).

Oppenheimer, M., and R. H. Boyle, *Dead Heat: The Race Against the Greenhouse Effect* (Basic Books, New York, 1990).

Park, C. C., *Acid Rain: Rhetoric and Reality* (Methuen, New York, 1987).

Parker, B., *Colliding Galaxies: The Universe in Turmoil* (Plenum Press, New York, 1990).

Parker, D., and J. Parker, *The Compleat Astrologer* (McGraw-Hill, New York, 1971).

Payne, J. B., *Encyclopedia of Biblical Prophecy* (Harper & Row, New York, 1973).

Perelman, L. J., *The Global Mind: Beyond the Limits to Growth* (Mason/Charter, New York, 1976).

Perrow, C., and M. F. Guillen, *The AIDS Disaster; The Failure of Organizations in New York and the Nation* (Yale University Press, New Haven, 1990).

Population Reference Bureau, *Population Handbook*, Second Edition (Population Reference Bureau, Washington, D. C., 1990).

Population Reference Bureau, *World Population Data Sheet* (Population Reference Bureau, Washington, D.C., 1990).

Redford, R., and T. J. Minger (ed.), *Greenhouse Glasnost* (Ecco Press, New York, 1990).

Ritchie, D., *Superquake: Why Earthquakes Occur and When the Big One Will Hit* (Crown, New York, 1988).

Roan, S. L., *Ozone Crisis* (Wiley, New York, 1989).

Robb, S., *Prophecies on World Events by Nostradamus* (Liveright, New York, 1961).

Ruff, H. J., *How to Prosper During the Coming Bad Years* (Warner Books, New York, 1979).

Sagan, C., and R. Turco, *A Path Where No Man Thought: Nuclear Winter and the End of the Arms Race* (Random House, New York, 1990).

Sande, M. A., and P. A. Volberding, *The Medical Management of AIDS*, Second Edition (W. B. Saunders, Philadelphia, 1990).

Schnarrs, S. P., *Megamistakes* (Macmillan, New York, 1989).

Schneider, S. H., *Global Warming* (Sierra Club Books, San Francisco, 1989).

Schneider, S., and R. Londer, *The Coevolution of Climate & Life* (Sierra Club Books, San Francisco, 1984).

Scientific American, *Earthquakes and Volcanoes* (W. H. Freeman, San Francisco, 1980).

Scientific American Readings, *The Science of AIDS* (W. H. Freeman, New York, 1989).

Simon, J. L., and H. Kahn (eds.), *The Resourceful Earth: A Response to Global 2000* (Basil Blackwell, New York, 1984).

Singer, F. S., *Global Climate Change* (Paragon House, New York, 1989).

Slaff, J. I., and J. K. Brubaker, *The AIDS Epidemic* (Warner Books, New York, 1985).

Smith, J. F., *The Coming Currency Collapse and What You Can Do About It* (Books in Focus, New York, 1980).

Sobel, R., *Panic on Wall Street* (Dutton, New York, 1968, 1988).

Steeger, W., and J. Bowermaster, *Saving the Earth* (Knopf, New York, 1990).

Thomas, K., *Religion and the Decline of Magic* (Scribner's, New York, 1971).

Thurber, J., *Fables for Our Time* (Harper & Brothers, New York, 1943).

Turner, R. H., J. M. Nigg, and D. H. Paz, *Waiting for Disaster* (University of California Press, Berkeley, 1986).

Velikovsky, I., *Worlds in Collision* (Macmillan, London, 1950).

Walters, L. M., L. Wilkins, and T. Walters (eds.), *Bad Tidings: Communication and Catastrophe* (Lawrence Erlenbaum, Hillsdale, New Jersey, 1989).

Weiner, J., *The Next 100 Years* (Bantam, New York, 1990).

Winch, D., *Malthus* (Oxford University Press, Oxford and New York, 1987).

Author Index

Subject Index